SEED TECHNOLOGY IN THE TROPICS

Other Related Books

Genetic Improvement of Field Crops	*Singh, C.B.*
Hand Book of Seed Industry (Prospects & its costing)	*Ramdeo, A.*
Handbook of Seed Science and Technology	*Basra, A.S.*
Identification of Weeds and their Control Measures	*Singh, Surjeet*
Organic Seed: Indigenous Varieties and Technologies	*Vanangamudi, K.*
Principles and Practices of Seed Storage	*Jusfice*
Seed Borne Diseases - Ecofriendly Management	*Arya, Arun*
Seed Borne Diseases Objectionable in Seed Production and their Management	*Bhale, M.S.*
Seed Economics: Commercial Considerations for Enterprise Management in Developing Countries	*Kugbei, S. (ICARDA)*
Seed Identification Manual	*Martin, A.C.*
Seed Pathology (Vol. 1-2 set)	*Neergaard, P.*
Seed Production of Agricultural Crops	*Kelly, A.*
Seed Quality Enhancement Principles and Practices	*Vanangamudi, K.*
Seed Science and Technology: Laboratory Manual	*McDonald, M.B.*
Seed Technology	*Khare, D.*
Weed Seed Biology	*Vanangamudi, K.*

SEED TECHNOLOGY IN THE TROPICS

Editors

D.B. MacKay
F. Ader
A.G. Gordon
C. Hutin
W.J. Rennie

Published by:
THE INTERNATIONAL SEED TESTING ASSOCIATION, ISTA

Reprinted by:
SCIENTIFIC PUBLISHERS (INDIA)

Reprinted in India, 2013
by
Scientific Publishers (India)

5-A, New Pali Road,
P.O. Box 91
Jodhpur - 342 001 (India)

4806/24, Ansari Road,
Daryaganj
New Delhi - 110 002 (India)

www.scientificpub.com

Seed Technology in the Tropics
Seed Science and Technology, Vol. 11/1, 1983

ISBN : 3-906549-31-3
ISSN : 0251 - 0952
First edition 1983

Print : 2022

Indian Edition ISBN : 978-81-7233-808-4

Editor : ISTA Editorial Committee, Chairperson : D.B. MacKay
Coordination : ISTA Secretariat, 8046 Zurich, CH-SWITZERLAND
The First reprint printed 1997 by :
Scientific Publishers (India), Jodhpur - 342 001, In-INDIA

Printed in India

CONTENTS

SEED TECHNOLOGY IN THE TROPICS

Butler, J. E. (1983), *Seed Sci. & Technol.*, **11**, 3–17

Purity testing in the tropics

J. E. BUTLER

Standards Branch, Department of Primary Industries, Brisbane, Australia

(Accepted October 1982)

Summary

Whilst most of the seed samples tested in the tropics are of species which are also widespread in temperate regions, substantial numbers of samples of tropical pasture species are also tested. Difficulties in the derivation and conduct of purity analyses of tropical species are discussed with particular emphasis on these pasture species. Methods and equipment requiring further investigation and research are highlighted.

Résumé

Analyse de pureté dans les tropiques

Bien que la plupart des échantillons de semences analysés dans les tropiques appartiennent à des espèces également répandues dans les régions tempérées, un nombre important d'échantillons d'espèces prairiales tropicales est également analysé. On discute les difficultés dans la mise au point et la conduite des analyses de pureté d'espèces tropicales en mettant l'accent surtout sur ces espèces prairiales. On souligne les méthodes et les matériels nécessitant des études et recherches complémentaires.

Zusammenfassung

Die Reinheitsprüfung in den Tropen

Während die meisten der in den Tropen geprüften Samenproben von Arten stammen, welche auch in den gemäßigten Regionen weit verbreitet sind, wird außerdem noch eine beträchtliche Anzahl von Proben tropischer Grasarten untersucht. Die Schwierigkeiten im Ursprung und in der Ausführung von Reinheitsanalysen tropischer Arten werden mit besonderer Betonung dieser Gräserarten diskutiert. Methoden und Ausrüstung zu weiterer Untersuchung und Forschung werden besonders hervorgehoben.

Introduction

The mention of the tropics conjures up images of wild, exotic places. Within the agricultural context, when one speaks of the tropics, one thinks of unusual species. Whilst it is true that many unusual species are grown in tropical agriculture, the development of agricultural and agricultural trade has not as yet been such that many of these species are involved in international trade to any significant extent, although several are quite promising (Innovation Board on Science and Technology, 1975).

Table 1. Species which are widely tested in tropical countries.*

Species	Number of samples	As a percentage of samples tested throughout world
Zea mays L.	14593	28.3
Triticum aestivum L. emend. Fiori et Paol.	6719	4.9
Sorghum bicolor (L.) Moench	3151	73.7
Phaseolus vulgaris L.	2848	20.4
Oryza sativa L.	2724	47.2
Panicum maximum Jacq.	1654	98.0
Pisum sativum L. s. l.	1560	6.5
Brassica oleracea L.	1362	7.5
Lycopersicon lycopersicum (L.) Karst. ex Farw.	1273	19.2
Glycine max (L.) Merr.	1265	61.7
Gossypium spp. (L.)	1260	55.1
Chloris gayana Kunth	1009	100.0
Vigna unguiculata (L.) Walp.	1003	78.6

* Based on data of Pedersen (1975).

Tropical species are, by and large, limited to a subsistence type farming on small areas and tend not to appear in statistics based on total yields or area sown. Data on seed testing in the tropics are very minimal. In 1974, the Referee Testing Committee of the International Seed Testing Association obtained information from laboratories all over the world regarding the species and the number of samples of each tested (Pedersen, 1975). Whilst African laboratories were well represented, there were only a few Asian laboratories and only one South American laboratory included.

I have selected 15 laboratories generally located in tropical or sub-tropical areas from this survey (Appendix 1) to determine the species most commonly tested in such regions. The species which are tested in largest numbers by laboratories in tropical regions are those which are not solely restricted to the tropics, e.g. maize (*Zea mays* L.) is by far the species of which the greatest number of samples is tested followed by wheat (*Triticum aestivum* L. *emend.* Fiori et Paol.), sorghum (*Sorghum bicolor* (L.) Moench) and beans (*Phaseolus vulgaris* L.) (table 1). Of these four crops, only sorghum can claim to have characteristics which make it more favourable in the tropics. Indeed, the number of maize samples tested in these 15 laboratories constituted only 28.3% of samples of maize tested by laboratories in the survey.

Rice (*Oryza sativa* L.), very much considered a tropical crop, was fifth on the list. However, the non-inclusion of laboratories from south-east Asia contributed to this ranking. Other data involving laboratories in Sri Lanka and Thailand (R. L. Harty, personal communication) indicate a greater role for rice in tropical seed testing of species not considered uniquely tropical (table 2).

Thus if one were to afford space in this review to species according to quantity of seed produced or number of seed samples tested, much of the discussion would

Table 2. Numbers of seed samples of species tested in Sri Lanka and Thailand*.

Species	Sri Lanka	Thailand
Arachis hypogaea L.	200	100
Beta vulgaris L.	150	–
Brassica oleracea L.	165	–
Capsicum spp.	400	20
Daucus carota L.	100	–
Glycine max (L.) Merr.	600	250
Hibiscus esculentus L.	200	–
Lycopersicon lycopersicum (L.) Karst. ex Farw.	225	30
Momordica charantia L.	125	–
Oryza sativa L.	6000	300
Phaseolus aureus Roxb.	250	120
Phaseolus mungo L.	100	–
Phaseolus vulgaris L.	900	–
Raphanus sativus L.	300	50
Solanum melongena L.	100	–
Sorghum bicolor (L.) Moench	20	170
Vigna unguiculata (L.) Walp.	1050	50
Zea mays L.	300	200

*Based on data from R. L. Harty (personal communication).

be about species that grow not only in the tropics but also in temperate regions. However, this special series is aimed at the tropical regions and therefore the interest and concern is for species which are uniquely or very largely tropical and which have their own particular characteristics or problems.

Perusal of the data in table 1 shows two pasture grass species which are tested in large numbers and for which the tests in these 15 laboratories represent almost all of the tests on these two particular species: *Panicum maximum* Jacq. and Rhodes grass (*Chloris gayana* Kunth).

Many of the recognised purity testing difficulties in the tropics involve pasture grass species. This is confirmed by examining the reports of the ISTA Purity Committee and its working groups involving tropical and sub-tropical species (Everson, 1975, 1978; Easton, 1972, 1975a; Vennell, 1978; Landenmark, 1981).

The development of the seed industry in tropical countries has been documented in a few instances (Mehta, Douglas and Singh, 1973; Chaudry, 1978; Joshi, 1978). However, most of these descriptions concentrate on seed production and administration of seed schemes and very little is said about the development of seed testing facilities. Some background to seed testing has been recorded for Taiwan (Chang and Chen, 1962) and Zimbabwe (Hanssen, 1966). Indeed Pili-Sevilla (1981) described seed testing in the tropics as being in its infancy.

In 1971 an appraisal of seed testing laboratories in 78 developing countries was conducted (Feistritzer, 1975). Seed testing had reached an advanced level in very few countries relative to developed countries (table 3) with the testing of cereals being

Table 3. Comparison of seed testing facilities and expertise between developed and less developed countries over five types of species (from Feistritzer, 1975).

	Percentage of laboratories in country classification (LDC, DC) rated as being within proficiency category (A, B, C)					
	A		B		C	
	LDC	DC	LDC	DC	LDC	DC
cereals	21.8	100.0	35.9	0.0	42.3	0.0
grain legumes	5.1	100.0	29.5	0.0	65.4	0.0
industrial crops	7.7	100.0	37.2	0.0	55.1	0.0
pastures	2.6	100.0	10.3	0.0	87.2	0.0
vegetables	5.1	100.0	16.7	0.0	78.2	0.0

Category A – advanced level
B – fragmentary or pilot scale operation
C – no activity reported.
LDC – less developed country
DC – developed country

more advanced than for other types of crops. Testing of pasture species was least advanced. There have been a number of efforts to improve seed testing within tropical countries by ISTA and FAO by providing expert advice and technical publications. The basic requirements for a laboratory to handle 2000 to 5000 seed samples per year are described in this number by van der Burg, Bekendam, van Geffen and Heuver (1983).

The training of analysts has been afforded quite a lot of attention. ISTA has conducted courses in Africa, Asia and Latin America (Wold, 1981). Various countries, e.g. New Zealand, have provided foreign aid to develop seed industries in tropical countries and this aid has also included the training of seed analysts. Seed technologists from tropical countries have been able to attend study courses at universities in Scotland, New Zealand and United States. There is now a pool of well trained analysts and technologists, and therefore the overall standard of analytical work in tropical regions should improve.

The discussion to this point has covered matters of relevance to all aspects of seed testing in the tropics. When the focus is on purity testing, there are several aspects which warrant consideration:

(I) sampling and sub-sampling,
(II) seed identification,
(III) pure seed definitions,
(IV) rapid mechanical aids for separating pure seed from inert matter.

Sampling and sub-sampling

The problems of sampling a seed lot are universal. Unless the sampling procedure is carried out accurately, no valid conclusions can be drawn about the quality of the seed lot no matter how precise the analytical measurements.

Seeds lots of tropical pasture grass seed can be extremely heterogeneous (Prodonoff and Bean, 1969). This is often brought about by unfavourable, unreliable climatic conditions in seed production areas and a lack of expertise where these species are treated as 'catch crops' without the necessary infrastructure and industry stability. These act to produce crops which are not uniform in composition nor maturity (Strickland, 1971). Furthermore, some tropical pasture seeds have certain peculiarities which make cleaning more difficult by comparison with seeds of temperate species (Linnett, 1977).

All instruments used in sampling seeds must be capable of reaching all parts of the seed lot and removing representative portions of each part sampled (Carter, 1965a). Nearly all samples are obtained from bag and bulk lots by means of 'seed triers' designed to accommodate the various seed sizes. Three types of triers are in general use: 'Stick', 'Nobbe' and 'Thief'.

The 'Stick type' is a double tubed probe with corresponding slots in each tube and long enough to extend the full length of the bag. This type is fully inserted into the bag in the closed position, opened by twisting and exposing the slots and then closed by again twisting. The trier when withdrawn is full of seed obtained from various positions within the bag or bulk.

The 'Nobbe type' is a single tube with a bevelled end and long enough to reach past the centre of the bag. It can be used to collect a trierful of seed in each insertion or can be inclined and moved so that seed flows through the tube for collection outside the bag.

The 'Thief type' is of similar length to the 'Nobbe type' but has an open slit extending practically its full length. Most of the seed sample obtained with this instrument comes from just inside the bag (Carter, 1965a). It is widely used and more suitable equipment is required particularly for rice (Pili-Sevilla, 1981).

Some of the tropical pasture species have chaffy seed (e.g., Gamba grass (*Andropogon gayanus* Kunth), buffel grass (*Cenchrus ciliaris* L. – figures 1a and 1b) which does not flow freely. Conventional seed sampling probes do not work well on these species (Larsen, 1975). Other seed characteristics, e.g. hooked pods in *Stylosanthes humilis* H.B.K. and *S. hamata* (L.) Taub, also restrict independent seed movement. Samples must be drawn by hand after making a small L shaped opening in the bag with a sharp knife. There is an urgent need for the development of better sampling devices for these species.

Many methods have been used to obtain working samples from the submitted samples. These include:

(I) halving or quartering method,
(II) dipping with spoons or scoops,

Figure 1. (a) *Andropogon gayanus*

Figure 1. (b) *Cenchrus ciliaris*

Figure 2. Soil divider

(III) random cups method, and
(IV) use of various mechanical dividers.

A good mechanical divider is the best way of obtaining a representative sample although the random cups method can be used with confidence (Carter, 1965b). The halving method or the spoon method are not recommended (Carter, 1965b).

The Boerner and Gamet dividers have been very popular with the Gamet yielding a slightly more accurate sub-sample (Hardin, Copeland and Knudsen, 1965). In the tropics, the soil divider (figure 2) has achieved great acceptance.

Seed dividers were designed to handle free-flowing seeds. With cnaffy grasses, dividers act on groups of seeds and hence are not suitable (Larsen, 1975). A modified hand quartering method is used in Queensland on *Cenchrus ciliaris* in which the seed mass is divided into eight approximately equal portions. Each alternate portion is discarded and the remaining portions remixed (figures 3a and 3b). This procedure is repeated until only the required amount of seed remains.

Leggatt (1949) indicated that a sample of approximately 3000 seeds was the size of sample which had been found as a result of extensive experience to be the most satisfactory in practice for determining pure seed percentage. The degree of accuracy is proportional to the square root of the numerical size of the sample and thus the gain in accuracy is reduced proportionately as the size of sample is increased. Miles (1963) calculated tolerances for purity based on 2300 chaffy seeds or 2750 non-chaffy seeds in working sample.

Figure 3. Hand Division of *Cenchrus ciliaris*. (a) division into eight

Figure 3. Hand Division of *Cenchrus ciliaris*. (b) removal of alternate eight

Thomson (1965) recommended that the difference between chaffy and non-chaffy seeds be ignored and that 2500 be accepted as the nominal minimum number of seeds for a purity test.

Felfoldi (1980) has published data on number of seeds per unit weight for a wide range of species. This enables the calculation of the required weights of seed for the various aspects of a purity test.

Papers have been published on working sample weights for purity analysis of *Corchorus capsularis* L. and *C. olitorius* L. (Verma and Arora, 1978) and *Eleusine coracana* (L.) Gaertn. (Arora and Banerjee, 1978). Vennell (1978) recommended working sample weights for many tropical pasture species.

Seed identification

Identification of weed seeds is a very important part of the purity analysis. Whilst seed identification techniques are similar in tropical and temperate regions, this task is made more difficult because of the lack of accurate seed reference collections. There is also an extreme shortage of books and references on seeds and their characteristics as contrasted to the availability of such texts for temperate species.

The problems are further compounded by the higher weed seed populations in tropical seed lots caused by inadequate management practices (Pili-Sevilla, 1981). Pasture species are once more the worst.

The presence of red rice in rice seeds has caused considerable consternation and been the subject of several papers (Sivasubramanian and Ramakrishnan, 1978; Hoagland and Paul, 1978; AOSA, 1980).

Pure seed definitions

ISTA has through the years developed definitions of pure seed which are universally applicable. These concepts apply just as much to the tropical species as to the temperate species.

The most basic concept is that pure seed shall be the unit of dissemination whether this unit be a true seed (in the botanical sense) or any one of a number of botanical structures such as achenes, schizocarps, mericarps (ISTA, 1976a). This concept can be applied to any new species for which one is seeking to devise a pure seed definition.

Difficulties can arise with the presence of appendages. Should the appendages be left attached or should they be removed and placed in the inert matter? There are examples in seed testing in which either of the approaches is used. The principle used to determine whether appendages should remain as part of the pure seed definition or be removed is whether the bulk of the seed sold commercially for planting has the appendage as part of the seed unit or whether the appendage is routinely removed during harvesting, cleaning or processing operations. If appendages are normally removed during such operations, then they are excluded from the pure seed definition.

Attached sterile florets are not included in the pure seed definitions (except for a few genera) and must therefore be removed (ISTA, 1976a, 1976b). Fortunately, some of the more important genera of the tropical pasture grasses (*Chloris* spp., *Panicum* spp., *Setaria* spp., *Sorghum* spp.) are included in the list of exceptions. Otherwise, a great deal of time and effort could be wasted.

By using the seed unit concept, the previously described principle for appendages and the 'half seed rule' (i.e. pieces of seeds more than half their original size are pure seed) a sensible pure seed definition for most species can be obtained.

Very great problems can arise when one has to distinguish or separate those seed units which satisfy the pure seed definition from those which do not. The legumes can be used to illustrate the difficulties which can occur. Pure seed is easy to define for large (e.g. peas (*Pisum sativum* L. *s.l.*)) or small (e.g. siratro (*Macroptilium atropurpureum* (D.C.) Urb.)) seeded legumes. However, when the seed unit consists of a seed plus pod material, the pure seed definition includes both the seed and pod segment. This is the case for *Stylosanthes* species such as *S. hamata, S. humilis* and *S. guianensis* (Aubl.) Sw.

Because of the high incidence of hardseededness within some tropical pasture legume species, mechanical scarification methods have been developed which, whilst reducing the level of hardseededness, also force some of the seeds from the pod material. International Rules preclude the use of pressure to determine whether an apparently intact pod contains a seed. There have been instances in which, although a large number of pods had lost their seeds during the mechanical scarification, it was not possible to tell which pods had lost the seeds without the use of manual pressure. Therefore, under International Rules these (empty) pods must be considered as pure seed. Pure seed percentage following scarification thus remains high. However, the germination of the seed lot can be dramatically reduced.

An inexperienced person might interpret the results as showing that mechanical scarification had reduced the quality of the seed lot to a considerable degree. Such a conclusion is incorrect if the primary effect of the scarification has been to remove seeds from pod material. As greater emphasis is placed on developing mechanical scarification procedures for legumes in which the seed unit is a pod or pod segment, this problem will become more severe. Many of the empty pods can be removed quite easily by the use of blowing techniques but these are not yet recognised for legumes.

We now turn our attention to the tropical grass species. The pure seed definition normally includes the presence of a caryopsis within the seed unit. The only exception is buffel grass in which the pure seed definition does not require the presence of a caryopsis before the fascicle is deemed to be pure seed (Cooper, 1981). For many of the grass species, the seed unit consists of a caryopsis enclosed within a lemma and palea and with subtending glumes (Strickland, 1971). It can be very difficult to readily determine whether or not such a seed unit contains a caryopsis (Prodonoff, 1967). We thus have the situation in which the pure seed definition requires the seed unit to contain a caryopsis to be classified as pure seed but the methods recognised by ISTA for the conduct of a purity analysis are not sufficient to enable the detection of a caryopsis.

There is therefore a need to bear in mind practical seed testing problems when fram-

Figure 4. General ER seed blower

ing pure seed definitions. For those species in which detection of caryopses is difficult, pure seed definitions might be framed so as not to require the presence of a caryopsis or mechanical separation techniques might be defined.

Mechanical aids

The two most widely used aids in purity testing for the separation of pure seed from inert matter are the seed blower and the diaphanoscope.

(I) *Seed blowers*

Seed blowers (figure 4) rely on a stream of air to remove light material and leave behind the heavier material in much the same way that 'winnowing' has been practised in agriculture for centuries. Seed blowers have to be calibrated so that the regulated air flow removes mainly those seeds units which do not contain a caryopsis but does not remove seed units which do contain a caryopsis.

There have been two approaches to the development of blowing procedures. One termed the uniform blowing procedure attempts to effect a separation such that the number of misplaced empty florets (i.e. empty florets left in the pure seed fraction) is equal to the number of misplaced full florets and that the number of these misplaced florets is small. This means that the pure seed percentage is close to that which would have been obtained had each seed unit been manually pressed to determine the presence of a caryopsis. The extent to which the numbers of misplaced florets vary from zero determines the extent of the effect on germination.

Since the pure seed percentage is as close as can be to the true purity as defined by the pure seed definition and since the germination of the pure seed must be reduced by the presence of misplaced empty florets, the pure live seed (P.L.S., i.e. the product of purity and germination) must be less than that which it is theoretically possible to obtain.

The other approach is the climax blowing method (Porter, 1940; Everson, 1957). This procedure relies on the fact that as the air speed is increased, the light (empty) florets are removed and they have a much greater impact on germination than on purity. Hence, P.L.S. is increased. As the air speed is raised further, the rate of change in P.L.S. slows until ultimately, the air is removing only pure seed thereby reducing the purity percentage without any effect on germination. P.L.S. then falls. The climax blowing method therefore utilises the blowing point which enables the maximum P.L.S. to be obtained by analysis. Not only do many firms trade in terms of P.L.S. as it provides a single figure for quality, but also it has been claimed that maximum P.L.S. represents the real planting value of seed (Porter and Leggatt, 1942; Leggatt, 1941; Everson, 1957).

Both the uniform and climax blowing methods are dependent on accurate and reliable calibration samples. Larsen (1981) has shown how to determine blowing points for several species using only one calibration sample. Calibration samples must be stored carefully and the calibration of blowers should be checked frequently.

Present seed blowers are more effective on free-flowing species. Species with awns and other appendages which limit independent seed movement tend not to be suitable for blowing although a blowing procedure has been approved for Rhodes grass (Easton, 1975b). The use of only small quantities of seed and location of stirrers in the blowing chamber to limit the 'matting effect' may enable some of the chaffy species to be blown (Easton, 1975a).

(II) *Diaphanoscope*

Whilst the use of diaphanoscopes is widespread throughout the world, they have a particular role in the tropics in the testing of grass seed by permitting the analyst to detect a caryopsis within the seed unit. Seeds are passed over a beam of light and the outline of any caryopsis is visible in quite a number of species. Thus, it is possible to ascertain whether a particular seed unit satisfies a pure seed definition requiring the presence of a caryopsis without damaging the seed unit or reducing its germinability (Justice, Musil, Andersen, Cull, Drake, Wertman, Caldwell, Kent and Zeleny, 1952).

Conclusion

Whilst a very large number of species is grown in the tropics, the majority of seed samples tested are of species which are also widely grown in temperate regions. Of the species which are predominantly tropical, those which are recognised as posing problems in purity analyses are pasture grasses.

Improved sampling and sub-sampling (dividing) equipment is required.

Pure seed definitions must be framed with a sense of realism. Where there are very great difficulties in determining whether or nor a seed unit complies with an existing pure seed definition, that definition should be modified. More use should be made of mechanical aids such as seed blowers and diaphanoscopes in the separation of pure seed from inert matter. As a consequence, greater effort is required to improve the design and functioning of these aids and in the standardisation of methods using them.

Whilst only a few short years ago, seed testing in the tropics was in its infancy, educational inputs in terms of regional workshops and specialised training have brought it to school age.

Appendix 1.
Location of laboratories from the survey of Pedersen (1975) used to derive data shown in table 1.

Country	*City*
Australia	Brisbane
Australia	Darwin
Chile	Santiago
India	New Delhi
India	New Delhi
Israel	Bet Dagan
Kenya	Nairobi
Malawi	Limbe
Mexico	Chapingo
Pakistan	Lahore
Phillipines	Manilla
Republic of South Africa	Pretoria
Tanzania	Arusha
Zambia	Chilanga
Zimbabwe	Salisbury

References

Arora, N. and Banerjee, S. K. (1978). Seed testing procedure for finger millet (*Eleusine coracana*). *Seed Research*, **6**, 158–160.

Association of Official Seed Analysts (1980). A potassium hydroxide test for the confirmation of red rice (*Oryza sativa* var.). *Newsletter of the Association of Official Seed Analysts*, **54**, 68–68.

Carter, A. S. (1965a). Seed sampling. In *International training course on seed improvement for Latin America. Campinas, Brazil. November 9–27, 1964* (ed. O. L. Justice and H. D. Bunch), pp. 22–23. International Seed Testing Association, Wageningen, Holland.

Carter, A. S. (1965b). Sample reduction. In *International training course on seed improvement for Latin America. Campinas, Brazil. November 9–27, 1964* (ed. O. L. Justice and H. D. Bunch), pp. 23–24. International Seed Testing Association, Wageningen, Holland.

Chang, H. S. and Chen, L. F. (1962). General introduction of the Taiwan Provincial Seed Testing Laboratory, Taichung, Taiwan. *Proceedings of the International Seed Testing Association*, **27**, 983–988.

Chaudry, A. R. (1978). Planning and implementation of seed production in Pakistan. In *Technology for increasing food production* (ed. J. C. Holmes), pp. 351–355. FAO, Rome.

Cooper, S. R. (1981). Report of the rules committee 1977–1980. *Seed Science and Technology*, **9**, 299–345.

Easton, G. R. (1972). Report of the working group on purity analysis of tropical seeds. *Proceedings of the International Seed Testing Association*, **37**, 331–337.

Easton, G. R. (1975a). Report of the working group on purity analysis of tropical seeds, 1971–1974. *Seed Science and Technology*, **3**, 85–87.

Easton, G. R. (1975b). Blowing procedure for purity analysis on Rhodes grass, (*Chloris gayana*). *Seed Science and Technology*, **3**, 511–514.

Everson, L. E. (1957). The climax blowing point for the analysis of Kentucky bluegrass. *Proceedings of the Association of Official Seed Analysts North America*, **47**, 73–78.

Everson, L. E. (1975). Report of the purity committee, 1971–1974. *Seed Science and Technology*, **3**, 81–84.

Everson, L. E. (1978). Report of the purity committee, 1974–1977. *Seed Science and Technology*, **6**, 141–144.

Feistritzer, W. P. (1975). World situation of seed production and supply. In *The role of seed science and technology in agricultural development* (eds. W. P. Feistritzer and H. Redl), pp. 37–45. FAO, Rome.

Felfoldi, E. M. (1980). *Seed counts (number of seeds per unit weight)*. Technical Report Series No. 32, Department of Agriculture, Melbourne.

Hanssen, K. B. (1966). Seed control in Rhodesia. *Proceedings of the International Seed Testing Association*, **31**, 243–249.

Hardin, E. E., Copeland, O. L. and Knudsen, L. A. (1965). A comparison of the relative effectiveness of the Boerner divider and several techniques of using the Gamet precision divider. *Proceedings of the Association of Official Seed Analysts North America*, **55**, 140–147.

Hoagland, R. E. and Paul, R. N. (1978). A comparative SEM study of red rice and several commercial rice (*Oryza sativa*) varieties. *Weed Science*, **26**, 619–629.

Innovation Board on Science and Technology (1975). *Underexploited tropical plants with promising economic value*. Report of an ad hoc panel of the Advisory Committee on Technology Innovation Board on Science and Technology for International Development Commission on International Relations. National Academy of Sciences, Washington, D.C.

International Seed Testing Association (1976a). International rules for seed testing. Rules 1976. *Seed Science and Technology*, **4**, 3–49.

International Seed Testing Association (1976b). International rules for seed testing. Annexes 1976. *Seed Science and Technology*, **4**, 51–177.

Joshi, M. S. (1978). The seed industry in Sudan. In *Technology for increasing food production* (ed. J. C. Holmes), pp. 351–355. FAO, Rome.

Justice, O. L., Musil, A. F., Andersen, A. M., Cull, H., Drake, V. C., Wertman, F. L., Caldwell,. B., Kent, C. A. and Zeleny, L. (1952). *Testing agricultural and vegetable seeds*. Handbook 30. USDA, Washington.

Landenmark, O. (1981). Report of the purity committee 1977–1980. *Seed Science and Technology*, **9**, 99–102.

Larsen, A. L. (1975). The testing of chaffy range grasses. *Newsletter of the Association of Official Seed Analysts North America*, **49**, 27–38.

Larsen, A. L. (1981). Use of a calibration sample of one species to determine the uniform blowing point of another. *Newsletter of the Association of Official Seed Analysts North America*, **55**, 26–29.

Leggatt, C. W. (1941). *The 'climax' blowing point in the testing of grass seeds for percentage of pure live seed*. Contribution No. 657, Division of Botany and Plant Pathology, Science Service, Department of Agriculture, Ottawa, Canada.

Leggatt, C. W. (1949). Sources of error in purity analysis. In *Handbook on Seed Testing*. Association of Official Seed Analysts of North America.

Linnett, B. (1977). Processing seeds of tropical pasture plants. *Seed Science and Technology*, 5, 199–224.

Mehta, Y. R., Douglas, J. E. and Singh, A. (1973). Historical development of seed improvement in India. *Seed Research*, **1**, 1–12.

Miles, S. R. (1963). Handbook of tolerances and of measures of precision for seed testing. *Proceedings of the International Seed Testing Association*, **28**, 525–686.

Pedersen, P. N. (1975). Report of the referee testing committee, 1971–1974. *Seed Science and Technology*. **3**, 280–337.

Pili-Sevilla, E. (1981). Tropical seeds in the Rules. *Fourth FAO/Norway workshop on seed testing for the tropics. Department of Agriculture, Bangkhen, Bangkok, Thailand. October 26–November 6, 1981.*

Porter, R. H. (1940). Our responsibilities as an association. *Proceedings of the Association of Official Seed Analysts of North America*, **32**, 62–69.

Porter, R. H. and Leggatt, C. W. (1942). A new concept of pure seed as applied to seed technology. *Scientific Agriculture*, **23**, 80–103.

Prodonoff, E. T. (1967). The determination and maintenance of seed quality. *Tropical Grasslands*, **1**, 91–98.

Prodonoff, E. T. and Bean, J. E. (1969). Heterogeneity in green panic seed. *Queensland Journal of Agricultural and Animal Sciences*, **26**, 215–223.

Sivasubramanian, S. and Ramakrishnan, V. (1978). Identification of rice varieties by laboratory techniques. *Seed Research*, **6**, 71–76.

Strickland, R. W. (1971). Seed production and testing problems in tropical and sub-tropical pasture species. *Proceedings of the International Seed Testing Association*, **36**, 189–199.

Thomson, J. R. (1965). Statistics committee 1962–1965. *Proceedings of the International Seed Testing Association*, **30**, 395–408.

Van der Burg, W. J., Bekendam, A., van Geffen, A, and Heuver, M. (1983). Project seed laboratory 2000–5000. *Seed Science and Technology*, **11**,

Vennell, G. R. (1978). Report of the purity committee working group on tropical crops 1974–1977. *Seed Science and Technology*, **6**, 145–149.

Verma, M. M. and Arora, N. (1978). Further studies on seed testing procedures for jute (*Corchorus capsularis* and *C. olitorius*) seeds. *Seed Research*, **6**, 151–157.

Wold, A. (1981). FAO/Norway workshops on seed testing for the tropics. *Seed Science and Technology*, **9**, 519–525.

Gregg, B, R. (1983), *Seed Sci. & Technol.*, **11**, 19–39

Seed processing in the tropics

B. R. GREGG*

Mississippi State University, Seed Technology Laboratory, Mississippi, USA

(Accepted July 1982)

Summary

Requirements and experience of processing seed in the tropics are reviewed. Principles of separating undesirable particles from seed lots are reviewed, and the functions of separators and other equipment used are described. Operating and management problems of processing seed in tropical areas are described, as are suggested solutions. Suggested processing sequences for some tropical seed crops are diagrammed.

Résumé

Conditionnement des semences dans les régions tropicales

Les exigences et les expériences de conditionnement des semences dans les régions tropicales sont passées en revue. Les principes de séparation des particules indésirables des lots de semences sont examinés et les fonctions des séparateurs et autres équipements utilisés sont décrites. Les problèmes de gestion et de mise en œuvre du conditionnement des semences dans les régions tropicales sont décrits et des solutions sont suggérées. Les séquences de conditionnement proposées pour quelques espèces de semences tropicales sont présentées sous forme de diagrammes.

Zusammenfassung

Saatgutaufbereitung in den Tropen

(Angenommen im Juli 1982)

Es wird ein Überblick über die Erfordernisse und Erfahrungen bei der Saatgutaufbereitung in den Tropen und ebenso über die Grundsätze der Abtrennung unerwünschter Bestandteile aus den Saatgutpartien gegeben. Ferner werden die Funktionen der Separatoren und weiterer Ausrüstungsgegenstände beschrieben. Betriebs- und Leitungsprobleme der Saatgutaufbereitung in tropischen Gebieten werden dargelegt und Lösungen dazu vorgeschlagen. Vorschläge zu Aufbereitungsanordnungen für einige tropische Saatgutarten werden in Diagrammen dargestellt.

Introduction

As harvested, most seed lots contain too much undesirable material, and must be processed to remove:

* Present address: Seed Division, Department of Agricultural Extension, Ministry of Agriculture and Cooperatives, 2143/1 Paholyothin Road, Bangkhen, Bangkok, Thailand

Weed seed
Inert material (trash, dust, hulls, leaves, stems, dirt, mudballs, etc.)
Other crop seed
Other cultivar seed (when mechanical separation is possible)
Undesirable seed of the seed crop (broken, immature, shriveled, insect-damaged, moisture-damaged, mold-damaged, diseased, etc.)

Seed conditioning (processing) is the seed programme component which removes undesirable materials, upgrades seed quality, treats seed with fungicides and insecticides, and measures it into desired packages. Processing is an essential part of tropical programmes which supply seed to small farmers.

Processing is a sequence of mechanical operations, any usually differs less between tropical and temperate zones than do production and marketing. Many major tropical crops are also grown in temperate zones; processing differences are minor and primarily due to local production conditions and specific requirements of different cultivars. The same separating principles and machines are used.

Seed processing

Seed processing includes all operations which prepare harvested seed for planting: handling, conveying, elevating, shelling, pre-conditioning, drying, cleaning, separating, sizing, upgrading, treating and packaging. A complete sequence of operations must be done in the proper order and at the proper time. Quality of the final seed lot depends on the thoroughness of removing contaminants and low-quality seed, size-grading for precision planting, treating the seed and preventing mechanical contamination.

Tropical seed processing has special requirements, critical both to achieving the desired seed purity and operating efficiency, and to maintaining seed viability. The tropical processor must strive for:

(1) *Complete separation* of undesirable material;
(2) *Minimum seed loss:* some good crop seed are removed along with contaminants in most separating operations; this loss should be minimised;
(3) *Upgrading quality* by removing rotted, cracked, broken, insect-damaged, injured and low-quality crop seed;
(4) *Efficiency:* the highest capacity consistent with effective separation; and at minimum cost.
(5) *Preventing loss of quality* by drying and processing seed before unfavourable conditions cause deterioration.

Seed conditioning includes several distinct steps which must be done in a specific sequence. First is RECEIVING, when seed arrives at the plant. After weighing, sampling and labelling, DRYING may be necessary to remove excess seed moisture. Seed may go into PRE-CLEANING STORAGE to be held for later processing, or may go directly to cleaning.

CONDITIONING AND PRE-CLEANING include removal of seed appendages and trash, hulling seed, shelling corn, etc., to permit efficient cleaning.

The first separating operation is BASIC CLEANING. The air-screen cleaner is the most common basic cleaner. Carefully-harvested seed lots from clean fields may be in good condition with few contaminants and require only cleaning on the air-screen cleaner. More often, special SEPARATING or UPGRADING machines must then be used to remove specific contaminants. These separate by differences in specific physical characteristics.

When all undesirable material has been removed, seed is ready for BAGGING. Sometimes a fungicide and/or insecticide TREATMENT is applied before bagging. Seed may require a second DRYING after treating, especially if it is sealed in vapourproof bags. Seed may then be SHIPPED directly, or held in STORAGE until needed.

Separating principles and equipment

Various seed processing machines separate by differences between good seed and undesirable particles in SIZE, LENGTH, SHAPE, WEIGHT, SURFACE TEXTURE, COLOUR, AFFINITY FOR LIQUIDS, or ELECTRICAL CONDUCTIVITY.

A single machine cannot separate by all physical characteristics; a different machine usually must be used to separate by each. Complete seed conditioning usually requires passing the seed through several machines in a specific sequence, each removing a portion of the undesirable materials.

Physical characteristics and machines which use them to separate undesirable materials from crop seed include:

PHYSICAL CHARACTERISTIC	MACHINE
SIZE	Air-screen cleaner Screen separators Width-thickness graders
WEIGHT (SPECIFIC GRAVITY)	Gravity separator Airblast of air-screen cleaner Aspirator Fractionating aspirator Pneumatic separator Stoner
LENGTH	Disc separator Cylinder separator Combined disc-cylinder separators
SHAPE	Spiral separator Draper belt
SURFACE TEXTURE	Roll mill (Magnetic separator) (Buckhorn separator)
COLOUR	Electronic colour sorter
AFFINITY FOR LIQUIDS	Magnetic separator Buckhorn separator

ELECTRICAL CONDUCTIVITY	Electrostatic separator

Other less common separators include the 'pin cylinder', which removes weevil-damaged peas by snagging the weevil hole with a pin; electromagnetically or mechanically vibrating surfaces; timothy bumper mill; etc.

To separate seed efficiently and effectively, the proper machines must be used; they must be set up properly, adjusted correctly and used in the proper sequence.

To achieve similar separations and seed quality, tropical seed programmes use processing machines and flow sequences very similar to those used in temperate zones. Tropical programmes which carefully control seed production (clean fields, careful roguing, isolation, insect and weed control, etc.) generally require processing only to upgrade seed purity and quality by removing inert material, damaged crop seed, and weed/other crop seed which are more economical to remove in processing than in the field. Most commonly used are the air-screen cleaner and gravity separator, with a length separator for rice and wheat seed.

Other equipment

Other equipment is required for efficient processing and handling of seed.
Also used are:

OPERATION	MACHINE	USE
Pre-conditioning	Scalper	Remove very large and small trash, dust, etc., from seed lots before drying, processing, etc.
	Debearder (hammermill sometimes used)	Remove awns, appendages etc., and improve test weight and flowability of grass and similar seed.
	Corn sheller	Remove maize seed from ears.
	Huller-scarifier	Remove small legume and similar seed from hulls and pods; scarify hard seed.
Drying	Dryer system (many types and methods of holding seed; fans; and burners to heat drying air)	Dry seed to safe moisture levels.
Polishing	Polisher	Improve appearance and brightness of large legume seed.
Hand-picking	Picking and sorting belts	Remove by hand specific undesirable materials (including maize ears) not removed by separating and cleaning machines.
Treating	Treater	Apply powder, liquid or slurry fungicides and/or insecticides to seed.
Weighing & bagging	Baggerweigher, bagger, packagers, scales (many types)	Weigh desired amount of seed into the desired package.
Bag closing	Bag closing system (many types: bag sewing machine, heat sealer, etc.)	Close and/or seal the filled seed package.

Elevating	Elevators (many types)	Elevate seed (usually vertically) for discharge into bins, etc.
Conveying	Conveyors (many types: for bulk seed, ear corn, bagged seed, etc.)	Convey bulk seed, ear corn, peanuts, bagged seed, etc., horizontally or at an angle.

Supporting equipment

Especially in the tropics, no seed conditioning plant is a separate entity which only conditions seed; most are largely self-supporting and carry out other activities to support processing. These essential supporting activities require special equipment which must be included in planning, budgeting and implementing the seed processing operation. Supporting equipment needed by seed conditioning plants includes:

1. Harvesting
   (Many tropical seed processing programmes must provide equipment for small growers to harvest/thresh seed)
   A. Dry seed harvest and threshing equipment
   B. Wet vegetable seed extractors

2. Sampling
   A. Triers and probes (bag, bulk, bin, etc.)
   B. Sample containers
   C. Hand scoops; samplers for flowing seed
   D. Sample storage and filing equipment

3. Moisture testing
   A. Moisture testers
   B. Ovens
   C. Balances and scales
   D. Sample containers

4. Purity testing, sample examination, etc.
   A. Forceps
   B. Purity workboards, tables, etc.
   C. Sample containers
   D. Balances and scales
   E. Magnifiers, lenses, etc.
   F. Working lights
   G. 'Black light' tester
   H. Hand testing screens, model-size seed separators, rice sample millers, sample-size scarifier-hullers, etc.
   I. Sample dividers

J. Sample blowers and aspirators
K. Test weight apparatus

5. Viability and vigour testing
   A. Germinators, racks, shelves, planting beds, etc.
   B. Tetrazolium testing equipment
   C. Germination pans, trays, etc.
   D. Lighting (if required)
   E. Vigour test equipment
   F. Refrigerators for storage, chilling, etc.
   G. Seed counting boards; vacuum seed counters

6. Weighing
   A. Truck scales (weighbridge)
   B. Warehouse platform scales
   C. Portable platform scales
   D. Portable bulk scales
   E. Bagger-weighers
   F. Sample-weighing scales
   G. Balances

7. Labelling and identification
   A. Label stencil cutters
   B. Stencil inkbrushes and stampers
   C. Tag printers
   D. Stack card holders, chalkboards, etc.
   E. Tag fastening tools
   F. Stock status boards

8. Monitoring conditions (drying, storage, ambient, etc.)
   A. Suitable thermometers (laboratory, bin, dial, dryer, etc.)
   B. Sling psychrometers; RH indicators (hygrometers, etc.)
   C. Recording hygrothermographs

9. Storage environment control
   A. Cooling and air conditioning equipment
   B. Dehumidifiers (desiccant and refrigerant)

10. Seed handling
    A. Seed/grain scoop shovels; 'scoop trucks'
    B. Bag holders
    C. Pallets (for forklift and/or stacking)
    D. Bulk seed handling boxes ('toteboxes', etc.)

E. Hand scoops
F. Bag trucks, carts, etc., to move bags
G. Forklifts
H. Hoists to dump bulk seed from motor trucks
I. 'Truck loading plates' to run bag trucks onto motor trucks
J. Bin level controls and switches

11. Clean-up and sanitation
A. Vacuum cleaner with hoses and nozzles
B. Air compressor with hoses and nozzles; hand-held blowers
C. Brooms and brushes
D. Rubber mallets, icepicks, etc.
E. Screenings, trash and waste product collection, handling, and disposal equipment
F. Bag cleaners

12. Dust and air liftings control
A. Dust and fume exhausters and hoods
B. Dust and light waste ducting and collection systems, handling and disposal equipment
C. Ventilation and air circulation fans

13. Insect control
A. Insecticide foggers (aerosol generators)
B. Sprayers

14. Pest control
A. Bait stations
B. Traps
C. Portable 'black light'
D. Utrasonic rodent and bird repellers

15. Fumigation
A. Gasproof fumigation covers
B. Gas detectors and concentration testers
C. Weights to seal edges of fumigation covers

16. Personnel safety
A. Gas masks and canisters suitable for fumigation gases
B. Dust masks, respirators and filters
C. Safety hats, goggles, gloves, etc.
D. Access ladders, work platforms, guard rails, etc.

17. Maintenance and tools
   A. Complete sets of electrician's, carpenter's, and mechanic's tools
   B. Drills, grinders and saws (hand; power portable and stationary)
   C. Arc and acetylene welding equipment
   D. Sledge hammers, crowbars, prybars, shovels, rakes, grass cutters, etc.
   E. Chain hoists, hydraulic jacks, and other equipment to move machines
   F. Tool cabinets and boxes, parts, storage bins, etc.
   G. Lubrication and greasing equipment
   H. Sheet metal-working tools and equipment
   I. Paint sprayer

18. Emergency equipment
   A. Electricity generators
   B. Emergency lights
   C. Firefighting equipment
   D. First aid kits, emergency oxygen, stretcher, etc.

Special requirements in the tropics

Tropical seed processing must implement each step in the proper sequence and at the proper time, manage them to improve seed quality, operate efficiently with the least cost and time, get seed ready for marketing at the necessary time, without delays which cause loss of seed or seed quality. Detailed requirements depend on local conditions, and often result in management requirements and emphases which differ considerably from those of temperate zones. The following have been major considerations in different tropical seed processing operations.

Cropping patterns

Many tropical areas can produce two or more crops each year. As seed storage is often difficult, especially for the large quantities required by farmers, it is more economic to plant seed in the season immediately following harvest than to 'carry over' to the next planting season. This complicates seed processing, as there may only be a short period between seed harvest and farmer planting time. This requires rapid harvest, drying and processing so that enough time remains to distribute and market the seed to farmers. This sometimes requires higher capacity processing facilities and/or more plants, as compared to temperate-zone operations with longer periods for processing and distribution.

Although seed processing is a seasonal operation, two daily operating shifts and operating seven days a week during short but critical periods are sometimes necessary in the tropics. Also, two major seed processing seasons sometimes occur each year; in some cases, processing goes on most of the year, with processing seasons of different crop seed overlapping.

To have seed ready when farmers need it, tropical seed production must be planned to allow as much time as possible, consistent with cropping patterns and weather, between seed harvest and farmer planting.

Production seasons
If irrigation or rainfall is adequate, tropical climates often permit planting seed crops during several periods, or in some cases at almost any time of the year. Seed crops may be grown in different seasons than commercial crops. With careful planning, seed crops can be planted at a time when moisture will be adequate to mature the seed fully, yet harvest and transport to the processing plant is during dry weather. This reduces seed drying requirements, deterioration which occurs under tropical conditions before seed can be safely dried, loss of seed lots, processing problems, and the amount of deteriorated seed which must be separated.

Receiving
Most tropical programmes either produce seed on large 'seed farms' within the programme, or under contract with small farmers trained as seed growers. The production system determines the method and facilities required to receive seed at the processing plant.

Large seed farms generally increase the cost of producing seed, due to the added cost of farming operations. However, seed operations can usually be controlled better and, when properly managed, result in high quality seed. As seed is harvested, it is delivered directly to the processing plant. Higher receiving and drying capacity is required, but seed can be properly dried and put in protective storage before significant deterioration occurs.

When small farmer-growers produce seed under contract, there is often delay in getting seed to the processing plant due to threshing methods, on-farm handling and purchase procedures. The processing plant requires less receiving-drying capacity, and operations are more geared toward bags. However, seed often suffer post-harvest deterioration due to delayed drying and poor handling on growers farms; viability and storability is reduced, and entire seed lots are sometimes lost.

In small-farmer tropical areas, the ideal blend of economics and technology is to produce seed with farmer-contract growers near the processing plant, assist growers with harvest so seed can be properly threshed at the proper stage, then immediately purchase seed and deliver it to the processing plant for proper drying and storage. This requires higher receiving-drying capacity for bagged seed or a special bulk delivery-handling system owned by the plant, equipment supplied by the plant and a procedure for immediate purchase of seed.

Drying
Tropical climates often have not only periods of heavy rainfall, but also high relative humidity, sometimes even during the dry season. Seed production must be planned so seed is harvested during low-rainfall periods, not only to avoid damage to seed,

but also to faciliate drying. Getting seed properly dried immediately after harvest is one of the most critical considerations of tropical seed processing.

Drying in the tropics is often more difficult because of high seed moisture at harvest, high relative humidity, and high ambient temperatures. Field and sun drying is useful but limited; artificial drying with heated air is required when RH is high; however, high ambient temperatures limit the increase in moisture-holding capacity of drying air heated to temperatures safe for seed. Drying thus may take longer, and a larger drying system capacity may be required.

Planning processing

Tropical seed processing operations must be carefully planned, because of the short time usually available for processing, necessity to dry seed properly under high ambient temperature and RH, delivery schedules and operating requirements. Drying must bring all seed to safe moisture levels before deterioration occurs. Seed to be sold first must be processed and ready for distribution when farmers need it. Seed for later sale can be dried and then stored unprocessed until seed for immediate sale has been processed.

Processing time can be reduced by careful management and planning, such as processing several lots of the same cultivar consecutively to reduce the number of clean-ups between lots of different cultivars. Lost time can be reduced by keeping machines operating during lunch hours, filling bins before the end of the work shift so that actual processing can begin early the next day, doing routine maintenance and clean-up during night hours when the plant is not operating, etc.

Storability tests are important in planning tropical processing. Much seed will be sold relatively soon, but some is often carried over to the following year. This is usually only six to nine months storage, but under tropical climates some crop seed will deteriorate during this period. To minimise losses, more storable lots should be identified and moved into good storage as soon as possible as a hedge in case they must be carried over. This is very important in many tropical seed programmes; as marketing may be subject to considerable fluctuation, there is often more carryover than was expected when the seed crop was planted.

Cleaning, separating and upgrading

Tropical seed fields have weed problems, including many found in temperate zones. Herbicides are less commonly used, but family and village labour is often cheaper and farms are smaller, so more seed field roguing can be done to remove weeds, off-types and volunteer crop plants. Traditional harvest methods also permit removal of many weed plants at harvest. However, weed seeds are fairly common in unprocessed seed lots of some crops such as rice (*Oryza sativa* L.), and proper processing is required to separate them.

Traditional harvest methods often leave more trash in the seed and may cause more mechanical damage. Closer cleaning with the air screen cleaner is required, cleaning capacity may be reduced and the percentage of weight loss in cleaning may be higher. On-farm winnowing of seed at harvest will remove much trash and dust and reduce the amount separated in processing.

However, it is safer to remove this material in processing and plan for the weight loss, as winnowing or hand cleaning on the farm delays delivery of seed to the processing plant for proper drying and storage.

Mudballs, clods, etc., from threshing floors are often found in tropical seed lots. While many of these are removed in basic cleaning, the gravity separator is often required.

Under tropical conditions, lightweight, immature, moisture- and insect-damaged seed and other light undesirable particles are more common in seed lots, often in higher percentages of the total weight. The gravity separator is often needed to bring such lots up to acceptable quality standards.

To make similar separations and process seed up to comparable standards, the same machines and separating sequences are required in the tropics as in temperate zones. In practice, however, few tropical processing operations use the more specialised separators to remove particular contaminants; tropical processing is more often limited to general cleaning and upgrading, because:

1. More low-cost hand labour is available, so field roguing and some hand separations are used instead of special separators.
2. Processing equipment is generally imported; equipment and spare parts procurement are often difficult and slow, and maintenance is less efficient. Thus, processing equipment is more basic.
3. Seed industries are not as highly developed, and do not serve commercial agricultures which demand extremely high quality seed. A lesser degree of separation and upgrading will provide seed which meets farmers demands, and is of better quality than other available seed.
4. A higher percentage of seed fields grown by farmer contract growers may be rejected.
5. Designers of processing plants are not as experienced; they do not prepare more modern layouts, but tend to provide simpler processing.

Processing sequences shown in Gregg (1977) are also suitable in the tropics. Additional sequences suggested for tropical seed crops, when processed in economic quantities, are shown in figures 1 to 14. In small quantities more hand cleaning and less mechanical processing is usual. Additional separators may be used to remove specific contaminants.

Corn (maize) (*Zea mays* L.) is commonly hand-sorted to remove undesirable ears before shelling. Seed is usually not closely size-graded in tropical seed programmes, as most small farmers plant by hand. Normally, cleaning and sizing with the air-screen cleaner and gravity separator are adequate for hand-planted seed. When farmers plant mechanically, a minimum number of size grades can be made. If seed volume is relatively low, gravity-grading to remove light particles can be done before the seed is size-graded. However, if seed volume is large enough, closer gravity-grading can be made after the seed has been size-graded.

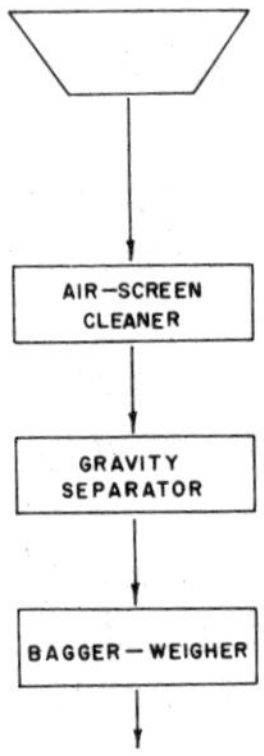

Figure 1. Simplified conditioning sequence for free-flowing tropical seed, without treating.

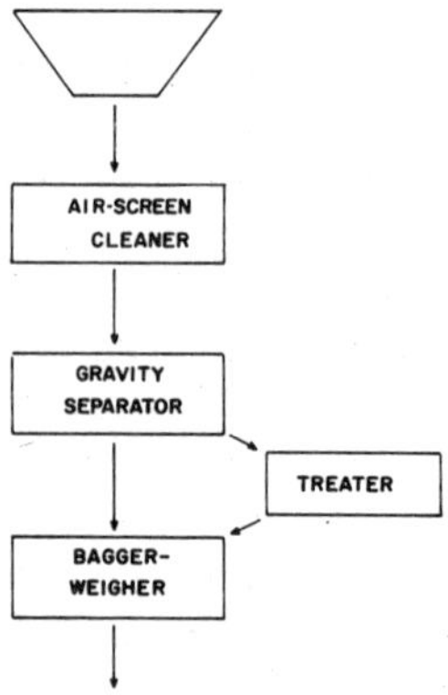

Figure 2. Suggested seed conditioning sequences for: Glycine (*Glycine javanica* L.) Flax (*Linum usitatissimum* L.) Kenaf (*Hibiscus cannabinus* L.) Rosella (*Hibiscus sabdariffa* L.) Sesbania (*Sesbania* spp.) Sesame (*Sesamum indicum* L.) Sunflower (*Helianthus* spp.) Safflower (*Carthamus tinctorius* L.) Moth bean (*Phaseolus aconitifolius,* Jacq.)

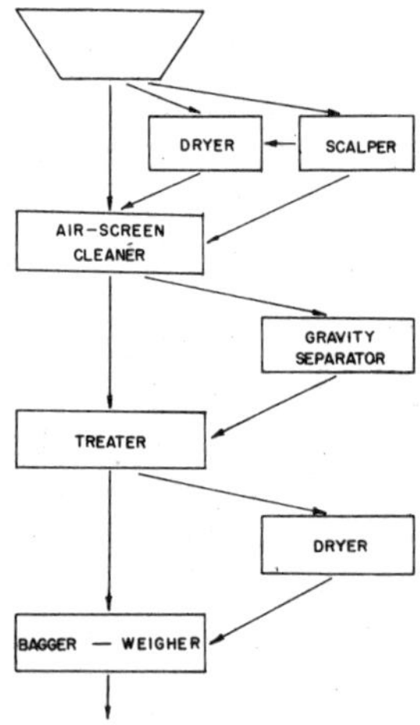

Figure 3. Suggested seed conditioning sequences for millets (*Setaria, Echinochloa, Pennisetum* spp.)

Rice (*O. sativa*) seed is commonly cleaned on the air-screen clearner alone, the air-screen cleaner followed by the gravity separator, or the air-screen cleaner followed by the gravity separator and a length separator, according to the materials which must be removed. Mudballs are common, as are light immature and unfilled seeds; such particles remaining after air-screen cleaning are largely removed by the gravity separator. Broken rice grains, short weed seed and some mudballs are removed by the length separator. While the gravity separator is not commonly used for rice seed in temperate areas, it is needed more often in the tropics.

Soybean (*Glycine max* (L.) Merr.) seed in the tropics presents more problems of immature seed or seed damaged by moisture, mold or threshing. Harvest at improper stages of maturity and poor post-harvest handling are common. Seed is commonly cleaned with the air-screen cleaner, followed by the gravity separator. Weeds are seldom a problem in seed from traditionally-harvested fields, as weed plants can be removed at harvest.

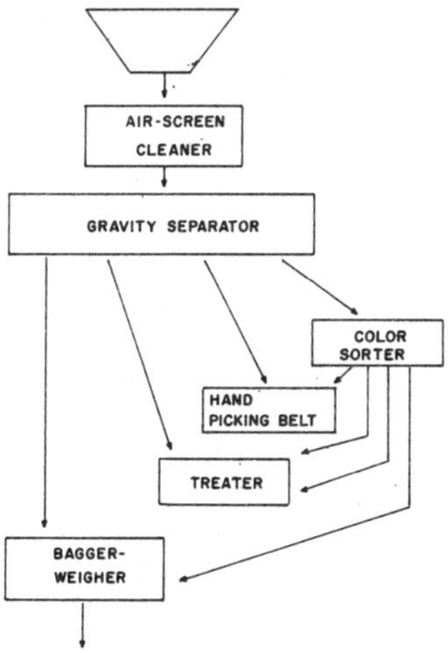

Figure 4. Suggested seed conditioning sequences for:
Adzuki bean (*Phaseolus angularis* (Willd.) Wight) Asparagus bean, yardlong bean (*Vigna sesquipedalis* Fruwirth) Broad bean (*Vicia faba* L.) French bean, haricot bean (*Phaseolus vulgaris* L.) Rice bean (*Phaseolus calcaratus* Roxb.) Scarlet runner bean (*Phaseolus coccineus,* L.) Lima bean (*Phaseolus lunatus* L.) Butterfly pea (*Clitoria ternatea* L.) Chickpea (*Cicer arietinum* L.) Cowpea (*Vigna unguiculata* (L.) Walp.) Lablab, Dolichos bean (*Dolichos lablab* L.) Mungbean (*Phaseolus mungo* L., *P. aureus* Roxb.) Velvet bean (*Mucuna* spp.) Lupins (*Lupinus* spp.) Pigeon pea (*Cajanus cajan* (L.) Huth) Lentils (*Lens culinaris* Medik.) Tepary bean (*Phaseolus acutifolius* A. Gray) Castor bean (*Ricinus communis* L.)

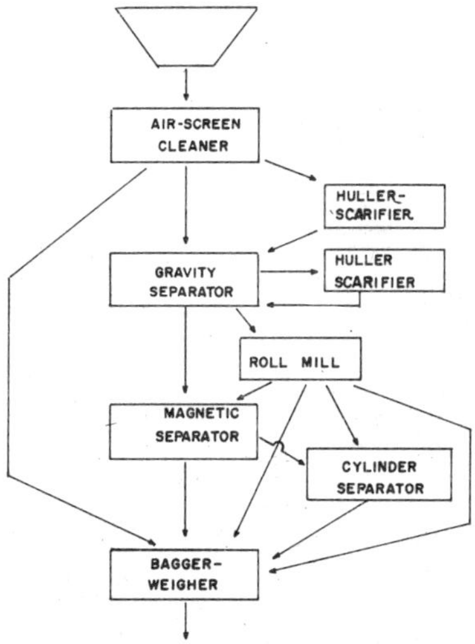

Figure 5. Suggested seed conditioning sequences for tropical forage legumes:
Black medic (*Medicago lupulina* L.) Alyce clover (*Alysicarpus vaginalis* (L.) DC.) Birdsfoot trefoil (*Lotus corniculatus* L.) Sweetclovers (*Melilotus* spp.) Centrosema (*Centrosema pubescens* Benth.) Clovers (*Trifolium* spp.) Lotuses (*Lotus* spp.) Crotalarias (*Crotalaria* spp.) Desmodiums, beggarweed (*Desmodium* spp.) Kudzu (*Pueraria lobata* (Willd.) Ohwil) Lespedezas (*Lespedeza* spp.) Medicks (*Medicago* spp.) Alfalfa, lucerne (*Medicago sativa* L.) Rough pea (*Lathyrus hirsutus* L.) Pueraria (*Pueraria* spp.) Sainfoin (*Onobrychis viciifolia* Scop.) Siratro (*Macroptilium atropurpureum* (DC.) Urb.) Stylosanthes (*Stylosanthes guianensis* (Aubl.) SW.), *S. humilis* H.B.K.)

Sorghum (*Sorghum bicolor* (L.) Moench) seed properly threshed can often be cleaned to the desired purity on the air-screen cleaner alone. However, the gravity separator is commonly used, to remove light materials and improve germination.

As peanuts (groundnut) (*Arachis hypogaea* L.) store best in the shell under tropical conditions, they are often not shelled until they reach the using farmer. Peanuts in the shell are normally processed over a peanut cleaner (a special type of air-screen cleaner) to remove dust, dirt, trash and some shells with immature light nuts. To improve quality, they are then cleaned over a gravity separator (which may be built into the peanut cleaner) to remove shells which are lightweight, empty or contain only

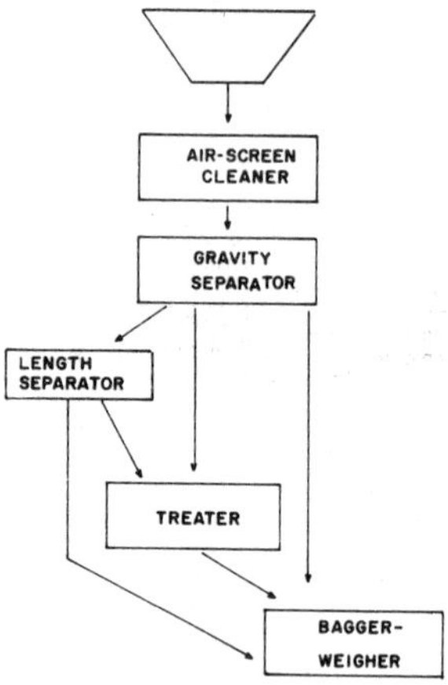

Figure 6. Suggested seed conditioning sequences for *Vicia* spp. (except *Vicia faba* L.)

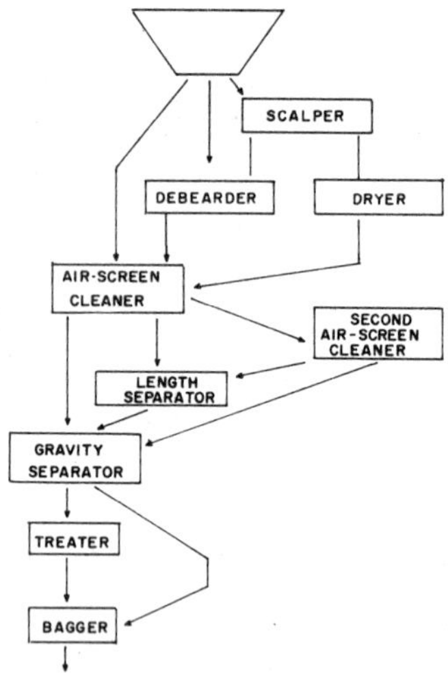

Figure 7. Suggested seed conditioning sequences for tropical grasses (*Gramineae* spp.)

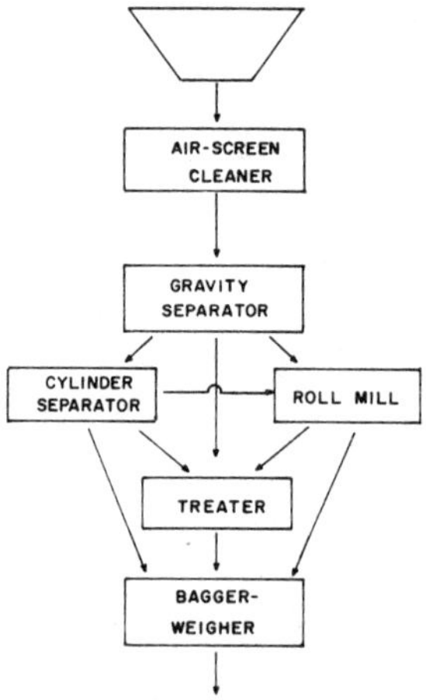

Figure 8. Suggested seed conditioning sequences for buckwheat (*Fagopyrum esculentum* Moench.)

small immature seed. Peanut seeds in the shell are not normally treated, as the using farmers normally shell them before planting.

Shelled peanut seed follows a processing sequence similar to that of temperate areas.

Where the output of processed seed can be handled without delay, the hand picking belt is useful for both shelled and unshelled peanuts, to remove offtype seed and undesirable particles remaining in the lot.

Cowpea (*Vigna unguiculata* (L.) Walp.) seed is usually cleaned on the air-screen cleaner, followed by the gravity separator to remove lightweight seed, heavy particles such as mudballs, etc., and to improve germination. The colour sorter or hand-picking belt is sometimes used to remove seed which are discoloured or do not fit varietal characteristics.

Seed of tropical grass species can generally be processed as shown in Gregg (1977).

Wheat (*Triticum* spp.) seed processing is similar to that of other areas. The common requirements are the air-screen cleaner, followed by a length separator to remove broken seed, short weed seed and particles of inert material. The gravity separator is sometimes used to remove light particles and mudballs remaining from traditional threshing. The spiral separator may be used to remove round weed seed; however, proper use of the air-screen cleaner and length separator will largely remove them.

Where wild oats (*Avena fatua* L.) are a problem, proper selection of seed fields can reduce its incidence. In cleaning, careful 'blanking' of the lower part of the first scalping screen of the air-screen cleaner, with heavy paper and tape, can remove most wild oats seed. The length separator can remove most of those remaining.

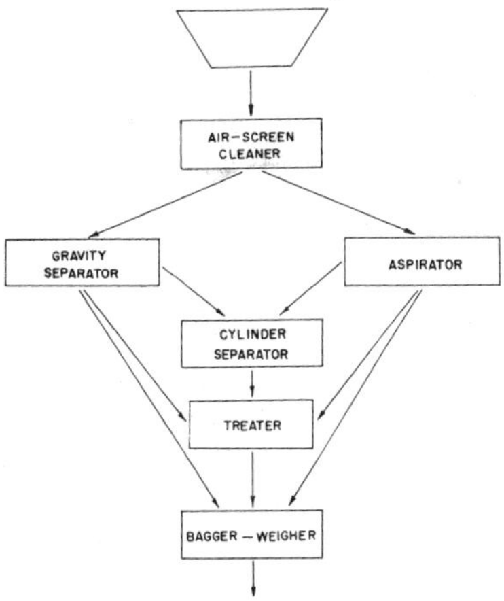

Figure 9. Suggested conditioning sequences for economic quantities of seed of *Brassica* spp. (broccoli, Brussels sprouts, cabbage, Chinese cabbage, cauliflower, collards, mustard, spinach mustard, pak-choi, pe-tsai, rape, rutabaga, turnip)

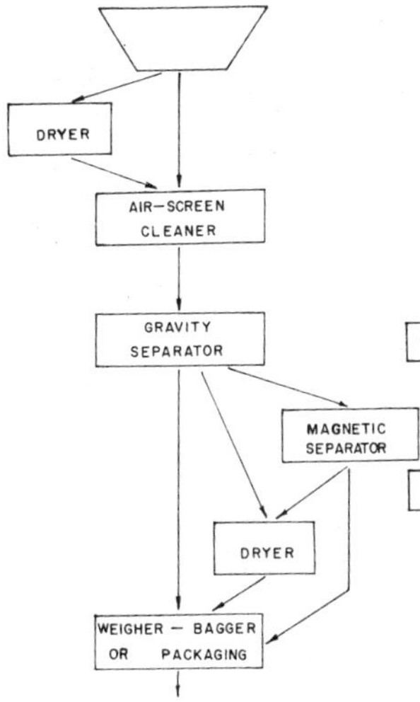

Figure 10. Suggested seed conditioning sequences for onion (*Allium cepa* L.)

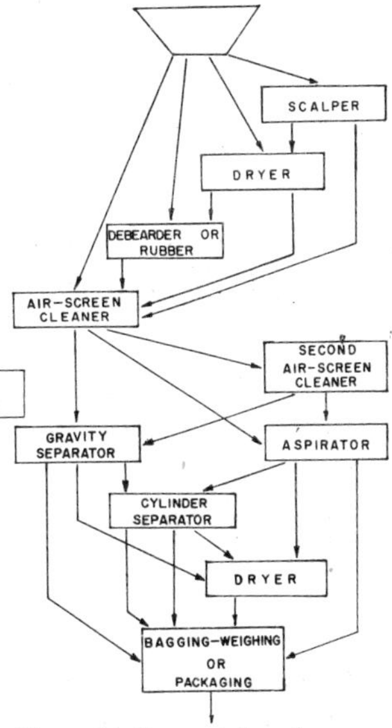

Figure 11. Suggested seed conditioning sequence for carrot (*Daucus carota* L.)

Mungbean (*Phaseolus aureus* Roxb.) and black gram (*P. mungo* L.) seed are commonly cleaned over the air-screen cleaner, and then on the gravity separator to remove light immature and insect-damaged seed, and mudballs if the seeds were threshed on a threshing floor. Although seldom used, a colour sorter would be useful to remove varietal mixtures. However, its low capacity and high cost generally make it more feasible in the tropics to rogue seed fields more carefully, and to accept some varietal mixture in the seed.

Beans (*Phaseolus* spp.) are usually well-processed. Primary cleaning is with the air-screen cleaner, followed by the gravity separator to remove light seed and particles and mudballs. The colour sorter and hand picking belt are used to remove offtype and discoloured seed.

Jute (*Corchorus* spp.) seeds are commonly cleaned with the air-screen cleaner, followed by the gravity separator to remove light seed and particles and heavier particles such as mudballs or sand. Traditionally-threshed jute seed lots often have small pieces of stems and similar inert material; higher seed purity is achieved with a disc separator to remove these particles. Some may be removed with a slotted wire mesh screen, but the screen perforations tend to 'plug' easily.

Elevators and conveyors

Despite low labour costs in many tropical areas, there must be complete uninterrupted bulk flow of seed through all processing machines from receiving to bagging. Thus, no delays or bottlenecks will occur in moving seed from one machine to another, while minimal investment and maintenance are required. The desired movement of crop seed can be handled safely and efficiently with properly-selected elevators and conveyors.

Some tropical processing operations use elevators and conveyors inadequately, under the impression that low labour costs make manual seed movement feasible between some machines. Such manual seed movement is, in most cases, ultimately more expensive than properly planned bulk flow: labour is less productive and long-term costs per kilogram are higher; seed movement is slowed, so total processing time is increased and seed delivery delayed; machines operate less efficiently and at erratic capacities, which sometimes affects processing efficiency; power costs are higher, as machines run more hours to process the same seed; labour is not uniformly productive at all times, and is sometimes not available as needed; and a full complement of workers is required when the plant operates, which reduces the total operating hours per day and output during peak processing seasons is less.

Determining separation requirements

Many plants have operated for years, and have well-trained and experienced operators and easily-accessible records of processing requirements of many lots. The processor thus knows or can quickly determine what must be done to process a given seed lot.

Some tropical processing plants do not have this backlog of experience and information; they may be relatively new, and their records and personnel have not yet reached the level which permits effective operation with minimum lost time and motion. Inadequate training and authority of operating staff, transfers or changes of personnel before their experience can be beneficial to the programme, and inadequate or poorly accessible records on previous processing reduce efficiency.

An inexperienced processor cannot simply move a seed lot into processing; this wastes time and seed, uses unnecessary machines, increases operating costs and seldom gives the best separation. To avoid this, tropical programmes should always determine the lot's processing requirements before beginning to process it. This is simple: a representative sample is separated, by hand as in a purity analysis, into:

A. Good seed which appears mature, full-size, healthy and undamaged.

B. Undesirable particles, weed and other crop seed, and damaged crop seed which should be removed.

Specimens of the good crop seed and undesirable particles to be removed are compared. Any physical characteristic in which they differ enough to make a separation is noted, and the available machine which can separate them is selected for use in the proper sequence.

Careful visual examination and comparison of seed, combined with experience and

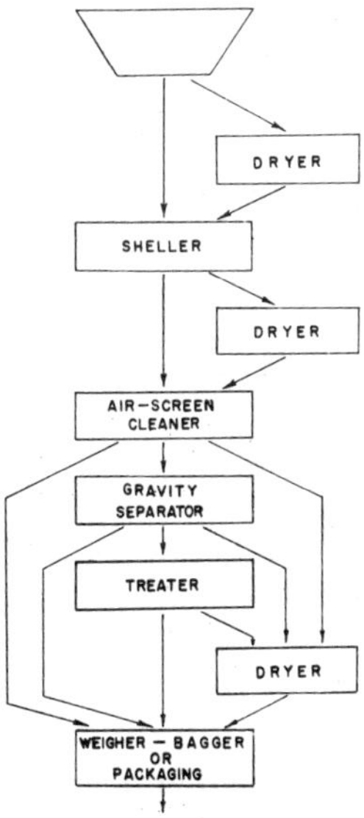

Figure 12. Suggested seed conditioning sequences for sweet corn (*Zea mays* L. var. *rugosa*)

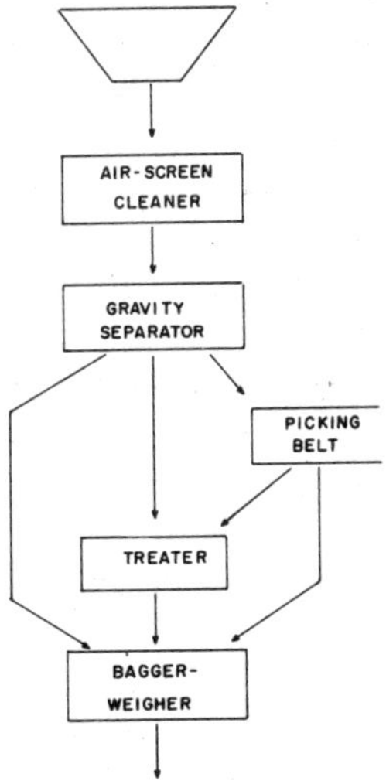

Figure 13. Suggested seed conditioning sequences for economic quantities of seed of cucurbits:
Citron, Ibara (*Citrullus lanatus* (Thunb.) Matsum. et Nakai) Cucumber (*Cucumis sativus* L.) Gherkin (*Cucumis anguria* L.) Gourds (*Lagenaria, Luffa, Telfairia, Benincasa, Cucurbita* spp.) Cantaloupe, muskmelon (*Cucumis melo* L.) Squash (*Cucurbita* spp.) Pumpkin (*Cucurbita* spp.) Watermelon (*Citrullus lanatus*, (Thumb.) Matsum. et Nak.)

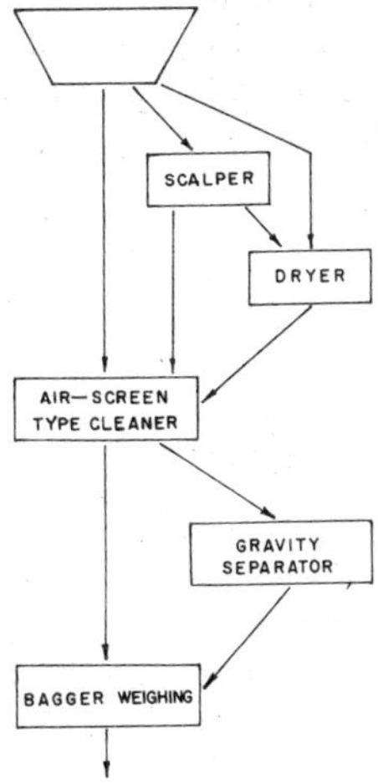

Figure 14. Suggested seed conditioning sequences for unshelled peanuts (*Arachis hypogaea* L.)

study, will indicate most separations and machines required. Some test machines and methods are available:

A. Screens for the air-screen cleaner: hand screens are used to select approximately correct screen sizes. Every tropical processing plant should have a complete set of hand screens.
B. Length separations: test model one-disc separator, or test model cylinder separator.
C. Surface texture separations: model roll mill.
D. Weight or specific gravity separations: small models of the gravity separator, fractionating aspirator and some other weight separators. A simple, inexpensive method of determining if the gravity separator can be used to separate seed is the waterglass (liquid sodium silicate) test.
E. Width-thickness separations: test model precision grader; seed can also be measured with a micrometer.

F. Shape separations; test model spiral separator.

G. Colour separations: the regular colour sorter can be used to predetermine separation efficiency and adjustments.

H. Affinity for liquids separations: test models of the magnetic separator; simple test methods to estimate differences in the ability of seed to absorb liquids and hold iron powder.

I. Conductivity separations: small test model electrostatic separator.

Most test models are expensive and more suited to research.

On most crop seed, approximate processing requirements can be determined simply by:

A. Analysing a sample of seed, to identify what must be separated and how it differs from the good crop seed.

B. Knowledge and experience of what each separator can do, how it does it, in what sequence it works best, and how to adjust and set it up to get the desired results.

C. Using hand screens to identify the approximately correct screen sizes.

Holding bins

Since all machines seldom operate at exactly the same capacity, a holding bin over each machine is necessary to collect and supply seed so the machines can operate uniformly at the best separation and capacity.

Holding bins are essential for efficiency in a tropical seed programme. A large bin over the air-screen cleaner permits workers to fill the bin quickly with enough seed for several hours of processing. The workers can then perform other duties such as bagging. With holding bins, less experienced operators can adjust each cleaner individually, without complicated balancing of capacity of machines. The treater can usually operate at a higher capacity; a large bin over the treater can accumulate seed, so it can be operated intermittently to save electricity and labour. If the bagging bin is large enough to hold the seed cleaned during several hours, seed can be bagged when workers finish other duties. Some tropical programmes use bagging bins large enough to hold all seed cleaned during one work shift; this shift operates with minimum personnel, and seed is bagged during the next shift by a larger work crew. This is especially useful during peak periods when night operations are required.

Bins allow processing for more hours daily; one or two persons can keep the machines operating during staggered lunch and 'break' periods. If the first bin is loaded with unprocessed seed at the end of the normal work shift, one or two persons can begin processing several hours early the next day, and gain several hours processing; seed can be bagged when the full work force arrives.

Forming lots

Tropical seed programmes which produce seed under contract with small farmers bulk together small amounts of seed from many farmers to form economic lot sizes. If seed quality is not fairly uniform among the growers, wide variation in seed quality among bags within a lot may result. Consequences can be serious, as a single bag often plants the entire crop of one small farmer.

To be sure that each bag of seed meets standards, variability among the 'sublots' put together to form a lot must be minimised. A procedure such as tetrazolium testing for rapidly determining quality of unprocessed seed identifies seed of similar quality to put together for more uniform lots. Within-lot blending, except as occurs during normal processing, is usually uneconomic due to cost of equipment and the large volumes of low-priced seed. Variability within lots may be reduced somewhat by using smaller lot sizes, especially when seed bags are small.

Packaging

Seed should be delivered to the small tropical farmer in a sealed, unopened bag, to minimise risk of contamination and to use the bag and label to promote improved seed. In many tropical areas, farms are small; seed bags should hold a smaller amount of seed, so that unopened bags can provide the correct amount of seed needed. This requires bagging seed in the amount used to plant a given area, usually related to the average field size of the smallest group of farmers using the seed.

Seed deteriorates faster when exposed to tropical conditions of high temperature and relative humidity during distribution and storage. To improve storability and minimise deterioration during distribution, there is increasing interest in packaging low-moisture seed in hermetically-sealed vapourproof bags. While this requires special drying, bags and sealing systems, as well as careful handling, it appears to offer solutions to seed deterioration under tropical conditions. It also adapts readily to programmes where bag weights are smaller, to provide unopened bags to small farmers with low seed requirements.

Handling bags

Handling bagged seed in small tropical programmes may lose much of the time gained by efficient processing. Bags may be handled with hand labour, in the belief that it is more economical. The most important consideration is to get seed to farmers on time; as more managers are realising, forklift and pallet handling of many small seed bags in a tropical seed programme is often more economical, and seed can be moved out faster. It also eliminates mechanical seed injury caused by workers dropping bags off stacks.

Trained personnel

Capable operating-level personnel, with the authorisations necessary to handle their responsibilities efficiently, are often lacking in tropical seed programmes. A seed programme requires a certain level of technology, regardless of local farmer technology; too often tropical seed programme personnel attempt to operate at the same level as local farmers. This has been ascribed to lack of adequate training, their agricultural experience with only local farmer technology, their difficulty in applying seed technology within a lower farmer technology context, and inadequate delegation of authority.

More advanced training is generally required for tropical seed programme personnel, to enable them to formulate, justify and implement operating solutions to seed production problems under tropical conditions with low farmer technology. While it would be excessive in a high-technology developed agriculture, the processing manager in a tropical programme can benefit from master's degree training in seed technology, combined with intensive practical training.

Constant training and skills upgrading for personnel down to the permanent worker level is essential. The person who actually does the job should be well-trained.

Every effort should be made to minimise turnover or transfer of trained personnel; the best training is properly-supervised practical experience, which is gained only over time. While periodic transfer of officers in a government seed programme is common, this requires a constant training investment to maintain competent staff, and the experience they gain is lost to the programme. Technical, managerial, operating and worker staff should be given every incentive to remain with the seed programme and develop their skills.

Insect control

Insect control is essential in tropical seed programmes, as seed often comes from the field with insect infestations.

As seed arrives at the processing plant, it should be examined for insects. As soon as possible, infested lots should be fumigated with a safe material such as hydrogen phosphide. Complete insect control should be practised, which includes keeping the entire area clean and free of trash, regular fumigation, treating seed with an insecticide before storage, spraying walls with a residual insecticide, and space sprays to kill flying insects.

Spare parts and maintenance

In tropical areas, seed conditioning equipment is often imported, as local equipment is not available or operates less efficiently. Inadequate spare parts are often purchased with the equipment, and are difficult to obtain later. To avoid excessive down-time due to mechanical breakdowns and lack of spare parts, tropical programmes should maintain large stocks of spare parts and practise good preventive maintenance.

Tropical temperatures and humidities cause faster deterioration of unprotected metal surfaces, rubber belts and flaps, etc. Careful, complete regular maintenance is necessary to minimise deterioration, replace worn parts, and maintain equipment. A qualified mechanic is essential, as most worker-operators in tropical seed programmes do not have the experience or training necessary.

Acknowledgements

Grateful acknowledgement is made to Opath Juta Sirivongse and Anusorn Paksansri for preparing the flow diagrams.

Sources of information

Agrawal, R. L. (1980). *Seed Technology*. Oxford and IBH. New Delhi, India.

Boyd, A. H. and Echandi, Z. R. (editors). (1978). *Seminario Internacional sobre Tecnologia de Semillas para Centroamerica, Panama, y el Caribe*. [International Seminar on Seed Technology for Central America, Panama, and the Caribean]. Universidad de Costa Rica y Laboratorio de Tecnologia de Semillas de La Universidad del Estado de Mississippi.

Copeland, L. O. (1976). *Principles of Seed Science and Technology*. Burgess. Minneapolis.

Delouche, J. C., and Potts, H. C. (1974). *Programa de Sementes: Planejamento e Implantação*. [Seed Programmes: Planning and Implementation]. AGIPLAN, Ministerio da Agricultura. Brasilia, D. F., Brazil.

Dougherty, G. M. (1982). *Seed drying and Processing: Corn and Sorghum. Proceedings, 13th Thai National Corn and Sorghum Conference*. Ministry of Agriculture and Cooperatives, Department of Agriculture. Bangkok.

Feistritzer, W. P. (editor). (1975). *Cereal Seed Technology*. FAO, Rome.

Food and Agriculture Organization (1961). *Agricultural and Horticultural Seeds*. FAO, Rome.

Gregg, B. R. (1977). Seed processing plant design with special reference to developing seed industries. *Seed Science and Technology*, **5**, 287–335.

Gregg, B. R. (1980). *Determining processing requirements*. DOAE Seed Technology Training Course. Phitsanulok, Thailand.

Gregg, B. R. (1980). *Seed processing*. 2nd Association of Southeast Asian Nations Seed Technology Workshop. Bangkok, Thailand.

Gregg, B. R., Law, A. G., Virdi, S. S. and Balis, J. S. (1970). *Seed Processing*. National Seeds Corporation. New Delhi, India.

Gregg, B. R. and Fagundes, S. R. F. (1975). *Manual de Operações da Mesa de Gravidade*. [Manual of Operation for the Gravity Separator]. AGIPLAN, Ministerio da Agricultura. Brasilia, D. F., Brazil.

Gregg, B. R., Wannapee, P. and Pluemsab, T. (1980). *Records in a seed program*. DOAE Seed Technology Training Course. Phitsanulok, Thailand.

Gregg, B. R., Wannapee, P., Pluemsab, T. and Saisavat, P. (1980). *The gravity separator*. DOAE Seed Technology Training Course. Phitsanuloke, Thailand.

Hawthorn, L. R., and Pollard, L. H. (1954). *Vegetable and Flower Seed Production*. Blakiston, New York.

Humphreys, L. R. (1979). *Tropical Pasture Seed Production*. FAO, Rome.

Jugenheimer, R. W. (1976). *Corn Improvement, Seed Production, and Uses*. Wiley-Interscience, New York.

Thomson, J. R. (1979). *An Introduction to Seed Technology*. John Wiley and Sons, New York.

Vaughan, C. E., Gregg, B. R., and Delouche, J. C. (1976). *Beneficiamento e Manuseio de Sementes*. [Seed Processing]. AGIPLAN, Ministerio da Agricultura. Brasilia, D. F., Brazil.

Harty, R. L, (1983), *Seed Sci. & Technol.*, **11**, 41–56

Testing of tropical species for germination

R. L. HARTY

Seed Testing Laboratory, Department of Primary Industries, Indooroopilly, 4068, Brisbane, Australia

(Accepted August 1982)

Summary

Except for tropical pasture species, germination testing in the tropics is dominated by plants of wide latitudinal distribution. This is due in part to the relatively small trade in agricultural seed of more uniquely tropical kinds. Exploitation of a wider range of useful species will depend upon the availability of practical information on production and testing techniques for use in seed quality assessment. Maintenance of adequate levels of germination is a particular problem in warm humid areas and vigour testing is of correspondingly less relevance although this is expected to change with improvements in seed storage. Lack of adequate laboratory facilities and suitably trained staff is occasionally reported but is probably a widespread phenomenon. Most of the utilised range of crop and pasture species are included in International Rules for Seed Testing but some modifications to the testing and reporting of dormancy in grasses and hard seed content in legumes are needed. Wider recognition of viability testing using tetrazolium would facilitate more rational testing procedures for tropical grasses.

Résumé

Analyse de la germination des espèces tropicales

Sauf pour les espèces fourragères tropicales, les essais de germination dans les tropiques sont dominés par des plantes de large distribution. Ceci est dû en partie au commerce relativement peu important de semences agricoles d'espèces uniquement tropicales. L'exploitation d'une gamme plus large d'espèces utiles dépendra de la disponibilité d'informations pratiques sur la production et sur les méthodes d'analyse à utiliser pour évaluer la qualité des semences. Le maintien de niveaux de germination corrects est un problème particulier dans les régions chaudes et humides et les essais de vigueur sont donc moins appropriés bien que ceci doive changer avec les améliorations dans la conservation des semences. On mentionne parfois le manque d'installations adéquates de laboratoire et de personnel convenablement formé, mais ceci est probablement un phénomène largement répandu. La plupart des espèces agricoles et fourragères utilisées sont incluses dans les Règles internationales pour les Essais de Semences mais il est nécessaire d'apporter quelques modifications dans l'analyse et l'indication des résultats de dormance chez les graminées et de pourcentage de graines dures chez les légumineuses. Une reconnaissance plus large des essais de viabilité au tétrazolium faciliterait des procédures d'analyse plus rationnelles pour les graminées tropicales.

Zusammenfassung

Die Prüfung tropischer Arten auf Keimfähigkeit

Mit Ausnahme von tropischen Graslandarten wird die Keimprüfung in den Tropen durch Pflanzen von sehr weiter Breitenverteilung beherrscht. Dies ist teilweise eine Folge des relativ kleinen Handels mit landwirtschaftlichem Saatgut von einzigdastehenden tropischen Arten. Die Ausnutzung einer größeren Reihe nützlicher Arten hängt ab von der Verfügbarkeit praktischer Informationen über Produktion und Prüfungstechniken zur Anwendung in der Qualitätsbeurteilung von Saatgut. Die Einhaltung entsprechender Keimfähigkeitshöhen ist ein besonderes Problem in warmfeuchten Gebieten und die Triebkrafprüfung ist von entsprechend geringerer Bedeutung, obwohl davon Änderungen bei der Verbesserung der Saatgutlagerung erwartet werden müssen. Über den Mangel an entsprechenden Labormöglichkeiten und geeignetem ausgebildetem Personal wird gelegentlich berichtet, das ist aber wahrscheinlich eine weitverbreitete Erscheinung. Die meisten der verwendeten Feld- und Weidearten sind in den Internationalen Vorschriften für Saatgutprüfung enthalten, aber es werden einige Abwandlungen in der Prüfung und Berichterstattung der Keimruhe bei Gräsern und des Anteils harter Samen bei Leguminosen benötigt. Eine weitergehende Anerkennung der Prüfung auf Lebensfähigkeit unter Anwendung von Tetrazolium würde rationellere Prüfungsverfahren für tropische Gräser erleichtern.

Introduction

This review will deal with the germination of important tropical field and plantation crop, fruit and pasture species, and allied subjects of dormancy, viability and vigour, drawing mainly on references selected from literature published between 1971 and 1981.

A survey conducted by the ISTA Germination Committee in 1981 disclosed that with the exception of pasture plants the major species components of tropical seed testing are not very different from those elsewhere. Seed testing in the tropical and temperate zones has surprisingly much in common. This is despite the fact that there is a great wealth of plant species utilised for food and other purposes in the tropics. Over 90 species were listed for tropical South America from ethnographic sources (Meighan, Pendergast, Swartz and Wissler, 1958) while across the Pacific a different suite of more than 100 species can be compiled for Java in the Republic of Indonesia (Mrs. T. S. Sudikno, personal communication).

About 3,000 plant species have been used for human food at various times throughout history and about 150 of these have been cultivated. Today most of the world is fed by about 20 species (National Academy of Sciences, 1975). In colonial times many indigenous species fell into disuse while others were prohibited for cultural or religious reasons and it has been claimed that it would be better to maintain and improve those food sources already present in less developed countries rather than to replace them with high technology-dependent Western crops (Marx, 1977).

Many locally utilised tropical species are in fact suitable for much wider use and examples of seed reproducing species in this category are listed in table 1. The wider exploitation of little used species in low latitudes is not likely to be a simple task. For a start, the tropics comprise approximately 38% of the land area of the world distributed between some 89 countries (Lugo and Brown, 1981). Again, extension of the species basis of agriculture involves not only technical considerations, but also a large range of socio-cultural systems (Gribbin, 1978; Bavappa and Jacob, 1981).

Table 1. Tropical plant species reproduced by seed, and suitable for wider exploitation*

Cereals	*Fruits*
Amaranthus spp.	*Annona muricata* L.
Chenopodium quinoa Willd.	*Citrus grandis* (L.) Osb.
Echinochloa turnerana (Domin) J. M. Black	*Durio zibethinus* Murr.
Zostera marina L.	*Garcinia mangostana* L.
	Guilielma gasipaes (Kunth) L. H. Bailey
	Pourouma cecropiaefolia Mart.
	Solanum quitoense Lam.
Vegetables	*Oil seeds*
Acrocomia spp.	*Caryocar braziliense* Cambess.
Bactris minor Jacq. Cogn.	*Cucurbita foetidissima* HBK
Benincasa hispida (Cogn.)	*Jessenia polycarpa* Karsten
Cocos nucifera L.	*Orbignya martiana* Barb. Rodr.
Euterpe oleracea Mart.	*Simmondsia chinensis* (Link) Schneider
Psophocarpus tetragonolobus (L.) DC.	
Forage	*Other*
Acacia albida Delile.	*Boehmeria nivea* (L.) Gaudich
Atriplex spp.	*Calathea lutea* (Aubl.) Schultes
Brosimum alicastrum Swartz	*Euphorbia antisyphilitica* Zucc.
Cassia sturtii R. Br.	*Mauritia flexuosa* L.
Prosopis tamarugo Phil.	*Paspalum varginatum* Swartz.

* from 'Unexploited Tropical Plants with Promising Economic Value' *Nat. Acad. Sciences, Washington D.C. (1975)*

The International Seed Testing Association has long recognised the need to develop germination testing methods for tropical species and thus assist in the development of agriculture in low latitudes (Meadly, 1972). Comparative germination testing has been a continuing task of the Tropical and Sub-Tropical Seeds Working Group of the Germination Committee and during the last decade 16 crop species, nine pasture legumes and eight pasture grasses were studied (Johnston, 1972, 1975a, 1978; Johnston and Harty, 1981).

Indigenous crops of the equatorial zone

Typical representative species of low-latitudinal agriculture are cacao (*Theobroma cacao* L.), coconut (*Cocos nucifera* L.), coffee (*Coffea* spp.), oil palm (*Elaeis guineensis* Jacq.), rubber (*Hevea brasiliensis* Willd. ex Adr. Juss. Muell. Arg.), sugar cane (*Saccharum officinale* L.), tapioca (*Manihot esculenta* Crantz.), tea (*Camellia sinensis* (L.) Kuntze), banana (*Musa* spp.) and pineapple (*Ananas comos* (L.) Merr.). Many tropical plants such as sugar cane, banana, pineapple are propagated vegetatively, and are no concern of seed testing except where seed is used in plant breeding programmes.

A nursery phase is a common feature of plantation crop production. Seedlings are raised in nurseries and transplanted to field plots often when quite large. With better chemical weed control it is possible that the raising of seedlings directly in the field may become a more economical operation (Williams, 1975).

Crops such as oil-palm and rubber are mainly propagated from seeds although in the latter case preference has been increasing for budding onto seedling rootstocks. Seeds must nevertheless be used to raise rootstocks and germinability of seed is important. An example of the difference between tropical and temperate agriculture can be seen in the practice of marketing germinated or pre-germinated seed of plantation crops by number. Orders are placed, not for weights of seed but for numbers of seed or seedlings (Chin and Lassim, 1980).

Germination of rubber seeds does not seem to present great problems but that of oil-palm is irregular and may spread over long periods of time (6–10 weeks) unless dormancy breaking procedures are adopted. Techniques of hydration and dehydration with heat (40°C) applied to seed in closed plastic bags have been described for large scale germination of oil-palm seeds (Noiret, 1968; Comont and Jacquemard, 1977). A relationship exists between heat requirement for dormancy breakdown and seed age (Odetola, 1974).

Because of the high commercial value of oil-palm seeds, destructive testing is uneconomic. Seed lots of about 1500 seeds are derived from individual bunches and Mok (1972a) suggested the grouping of bunches into batches of about 10,000 seeds. Samples of 400 seeds were found to provide statistically acceptable levels of precision for experimental work and commercial testing.

Patterns of germination of palm seeds have been described by Sento (1976) who also recommended the use of tetrazolium for viability testing of *Butia* and *Phoenix* spp. Good agreement between germination and tetrazolium-determined viability has also been reported for oil-palm seed (Mok, 1972b).

Germination behaviour of a wide range of crops has been described in the period under review: *Areca* sp. (Das and Baruah, 1974), *Bixa orellana* L. (Goldbach, 1979), *Coffea arabica* L. (Valio, 1976), *Corchorus* spp. (Singh, Karma and Verma, 1972; Verma and Arora, 1978; Okusanya, 1979), *Datura innoxia* Miller (Singh 1974), *Dioscorea* sp. (Tyagi, 1973), *Eleusine coracana* Gaertn. (Arora and Banerjee, 1978), *Hibiscus esculentus* L. (Omran, El-Bakry and Gawish, 1980), *Manihot esculenta* Grantz. (Ellis and Roberts, 1981). Physiological investigations of dormancy in African species of *Cucurbitaceae* were reported by Heit, Robinson and Mishanec, (1978): these tropical species exhibit marked dormancy characteristics in contrast to their temperature counterparts. Use of tetrazolium has been effective in assessing viability of dormant coffee seeds (Goldbach and Aguilera Vizcarra, 1980) and *Eleusine coracana* (Agrawal and Kaur, 1975). Dormancy in the latter species can be broken with KNO_3 or by scarification (Kulkarni and Basavaraju, 1976).

Seed germination is important as a method of increasing the food value of grains and there is a need for studies of sprouting techniques (National Academy of Sciences, 1974). Methods of producing edible sprouts have been described by Ruberte (1977).

Tropical crops of wide latitudinal distribution

Species of economic importance both inside and outside the tropics include cotton (*Gossypium* spp.), maize (*Zea mays* L.), peanut (*Arachis hypogaea* L.), rice (*Oryza sativa* L.), sorghum (*Sorghum bicolor* (L.) Moench), soybean (*Glycine max* Merr.) and sunflower (*Helianthus annuus* L.).

Dormancy effects

Rice is the major cereal crop of the tropics, where about nine-tenths of the world's rice is produced and consumed (Lassim and Chin, 1980). Storage is a major problem and control of moisture content vital. Considerable attention has been paid to dormancy in rice. Less dormancy and higher germination characterises the 'japonica' compared with 'indica' types (Hsieh, 1973). High storage temperature is effective in removing dormancy in rice seed (Yasue and Asia, 1970; Yasue, 1973). Various aspects of the physiology of rice seed germination in relation to light and oxygen have been investigated by Kordan (1976, 1977).

Factors affecting dormancy breaking in peanut seeds with increased storage temperatures were discussed by Sharir (1978). Dormancy in this species has been associated with endogenous ethylene (Ketring and Morgan, 1969) and may be broken with ethephon (Joshi, Mishra and Gaur, 1978). The implications of ethylene for seed testing are of particular interest. Ethylene is microbically produced in the soil by spore forming bacteria and precise manipulation of the oxygen ethylene cycle may be a future means of increasing productivity through minimum tillage and the use of organic amendments in farming (Smith and Cook, 1974).

Allied to this is the 'safe-sites' concept of Harper (1977) in which the number of seedlings which establish successfully is a measure of the 'safe-sites' which afford *inter alia* the stimuli to break dormancy. This would appear to provide a rationale for the use of ethylene in the laboratory testing of dormant seed.

Short-term dormancy of freshly harvested sorghum is relatively common and sometimes causes problems in breeding programmes. A wide range of dormancy-breaking procedures were investigated for the species (Gaber, Abdalla and Mahdy, 1974). Methods prescribed by the International Seed Testing Association (1976) were satisfactory combined with pre-chilling at 5 °C or 10 °C for up to seven days. Low temperature treatment was also the most effective remedy for dormancy in sunflower, either alone or in combination with gibberellic acid (Cseresnyes, 1979).

Seed vigour

A major difference between tropical and temperate cropping of similar species is the disproportionate amount of attention given to vigour testing between the two parts of the world. In the United States of America, vigour has been perceived as the most important seed quality factor to be pursued (Spain, 1976). Physiological quality was seen to be the most chronic problem facing the soybean (TeKrony, Egli and Balles, 1980) and cotton industries (Scott, 1980). No such emphasis can be found in the tropics probably because of the scope and dimension of more basic problems.

Soybeans would be a case in point. A relatively minor crop in the tropics (compared with rice) soybean expansion in Ghana has been inhibited by poor establishment (Nangju, Wien and Ndimande, 1980). Maintenance of viability is a more serious problem in the warm humid tropics, but with improvements in seed production techniques and seed storage facilities, the vigour issue will probably become important here as well.

Various vigour concepts have been expressed but few have been adequately examined by seed analysts even in the developed countries. Multi-criteria (Abdul-Baki and Anderson, 1973) and single criterion approaches (Roberts and Ellis, 1980) have been proposed. Accelerated ageing appears to hold the most promise for vigour testing of soybeans (TeKrony and Egli, 1977; Tao, 1978) or a modification of the accelerated ageing test in which seeds are aged at constant moisture content (Matthews, 1980). Measurement of conductivity of seed leachates has been found to accurately assess soybean seed quality affected by mechanical stress or ageing (Mc Donald and Wilson, 1980).

Cotton seed probably receives the harshest treatment of any seed because of its harvesting and delinting processes. A useful account of the sources of seedling abnormality has been provided by Gelmond (1979). Leachate conductivity has been used as a cotton seed vigour test (Hopper and Hinton, 1980). Low vigour has been attributed to partially filled seeds and flotation methods have been used to select vigorous planting material (Turner and Ferguson, 1972), a method also employed for similar reasons in rice (Kamil, 1977).

Slow emergence of non-dormant sorghum seed frequently occurs in early season sowings under cool conditions and a range of vigour tests has been examined for this species. Germination in the presence of ammonium chloride has been a stress test which at times has correlated well with field emergence (Souza and Marcus Filho, 1975) but there is no clear evidence of its superiority to standard germination tests (Yaycock, Jan and Vanderlip, 1975).

Despite the intensity of research into seed vigour it is unlikely that the standard germination test will be supplanted in the large scale screening of seeds for sowing quality (MacKay, 1978). Constructive progress in the development of useful alternative test procedures would be assisted by an apportionment of seed testing objectives and practices between the different requirements of seed as an article of trade and as material with specified physiological attributes (Ellis and Roberts, 1980).

Tropical fruit crops

Tropical dessert fruits are recognised to include some of the most luscious of all fruit. The durian (*Durio zibethinus* Murr.) is particularly well known for its strong, unusual flavour, and other highly esteemed species are the langsat (*Lansium domesticum* Correa), mangosteen (*Garcinia mangostana* L.) and rambutan (*Nephelium lappaceum* L.).

These and many other fruits are virtually unknown outside the tropics and would have good prospects for expanded production and marketing in bordering countries and ultimately in more distant latitudes. These species are usually reproduced from seeds although in some cases variable performance of progeny has called for grafting methods of reproduction (Chua and Teoh, 1973).

The seed of tropical fruit species is notoriously short lived which imposes constraints on expansion of production. Seed of the durian is sensitive to high temperatures and can be damaged even by a short exposure to direct sunlight (Malo and Martin, 1979). Viability can be maintained for about a month at 20 °C and about six days at 36 °C (Soepedmo and Eow, 1977). Seeds of durian have been successfully transported in plastic bags packed with slightly dampened paper for up to two weeks (Malo and Martin, 1979). Germination difficulties have been described for pawpaw (*Carica papaya* L.), durian, jackfruit (*Artocarpus integrifolia* L.), rambutan, mangosteen, mango (*Mangifera indica* L.) and *Citrus* species (Teng, 1980). Of interest is the finding that rambutan seeds maintained better germination in a medium moistened with the juice of their own arils which possesses germination inhibitory properties (Chin, 1975). Germination procedures have been described for soursop (*Annona muricata* L.) (Morton, 1966), langsat (Whitman, 1980), ber (*Zizyphus mauritiana* Lamb) (Kirpal Singh, Chadha and Gupta, 1973) and mango (Singh and Kaisuwan, 1971).

It is difficult to predict if testing seed of these species will be required to any great extent but it is certain that their germination characteristics need better documentation, and it is likely that some quality control will be necessary in time. There is a dearth of description in the literature of basic embryogeny and seedling morphology in relation to developmental stages of tropical fruit species and more specific documentation of seed and seedling characteristics should be encouraged. Seed testing laboratories could be appropriate centres for collating existing information and for initiating the required descriptive studies.

Pasture species

Sown pastures are important in the tropics because they facilitate the development of livestock farming and the production of food supplies from areas unsuitable for cropping. Nearly half the world's cattle population is found in the sub-humid tropics and it is these areas where the development of sown pastures has begun to take effect (Boonman, 1980). More widespread use of improved pasture cultivars depends on the availability of seed which will reliably establish good pastures when planted. Many tropical forage plants represent relatively wild species which have not been domesticated and rigorously selected for good seed production characteristics (Humphreys, 1979). Seed production and testing problems of grasses and legumes were reviewed by Strickland (1971).

Legumes

Of the thousands of known legume species in the world, most are of untested agricultural potential (National Academy of Sciences, 1979). Useful information on 146 legume species has recently been provided by Duke (1981). Seed production, certification and testing of the major tropical pasture legumes has been detailed by Humphreys (1979). Species described included *Centrosema pubescens* Benth., *Desmodium intortum* (Miller) Fawcett and Rendle, *Desmodium uncinatum* (Jacq.) DC., *Neonotonia wightii* (Arn.) Lackey, *Lablab purpureus* (L.) Sweet, *Leucaena leucocephala* (Lam.) de Wit, *Macroptilium atropurpureum* (DC.) Urban, *Pueraria phaseoloides* (Roxb.) Benth., *Stylosanthes guianensis* (Aubl.) Swartz, *Stylosanthes humilis* Kunth, and *Stylosanthes scabra* Vog..

These species present few problems in seed testing. Viability is usually high but a significant proportion of seed is impermeable to water. Other forms of dormancy are uncommon in genera other than *Stylosanthes,* and dormancy breaking procedures are not necessary in seed testing. Complex germination responses to temperature were first reported for *Stylosanthes humilis* by Ballard (1972). Embryo dormancy lasts up to six months, but germination was induced by single, brief exposures to small (2.5 °C–5 °C) temperature shifts or by treatment with thio-urea (Ballard and Buchwald, 1974). Alternating short-durational temperature cycles were recommended for testing this species (Butler, 1975).

Heating (85 °C) and cooling (ambient temperature) cycles were effective in reducing seed coat impermeability of *Stylosanthes humilis, Stylosanthes hamata* (L.) Taub., *Stylosanthes scabra* and *Stylosanthes villosa* Swartz (Mott and McKeon, 1979). Germination responses to temperature were described for seven species of *Stylosanthes* (McIvor, 1976).

Laboratory methods for overcoming hardseededness have also been described for a range of species (Phipps, 1973): *Centrosema pubescens, Phaseolus atropurpureus* DC. (*Macroptilium atropurpureum*), *Pueraria lobata* (Willd.) Ohwi, and *Stylosanthes gracilis* (*Stylosanthes guianensis*). Concentrated sulphuric acid scarification is most universally found to be effective in removing seed coat impermeability (Pe, Hill and Johnston, 1975; Febles and Padilla, 1977) but dry heat (Mott and McKeon, 1979) or hot water treatment (National Academy of Sciences, 1977) are sometimes recommended.

Opinion on the agronomic significance of hard seed is rather divided. In some species such as *Leucaena leucocephala* succesful plant establishment is necessarily dependent upon treatment of the impervious waxy seed coat before sowing (National Academy of Sciences, 1977) but this procedure has not been strongly recommended for other species. Pasture agronomists on one hand tend to regard high hard seed levels as necessary for the persistence of adequate populations of plants and the maintenance of yield (Jones and Evans, 1977; Humphreys, 1980). On the other hand rapid establishment is necessary in tropical pasture environments where weed competition is often very strong. High hard seed content often results in poor initial stands and, irrespective of the role of hard seed in the subsequent life of the pasture, there is a need for more research into practical methods of reducing hard seed levels in initial sowings.

Ballard (1972) referred to a basic inconsistency in seed testing on this matter. Whereas dormancy breaking procedures are permitted, no allowance is made for breaking

Table 2. Grass species of future importance for possible inclusion *in* International Rules for Seed Testing

Species	*Common name*
Andropogon gayanus Kunth	Gamba grass
Bothriochloa insculpta (A. Rich) A. Camus	Greeping bluegrass
Brachiaria humidicola (Rendle) Schweicherdt	Koronivea grass
Digitaria smutsii Stent	Tall finger grass
Hyparrhenia rufa (Nees) Stapf.	Thatch grass
Pennisetum clandestinum Chiov.	Kikuyu grass
Pennisetum purpureum Schumacher	Elephant grass
Setaria porphyrantha Stapf	Purple pigeon grass

of hardseededness. If all hard seed were viable there would of course be no point in testing viability, but levels of dead seed have been found approximating to 30% of the hard seed for a number of tropical legume species (J. E. Butler, personal communication). There is need for modifications to reporting procedures for viability of tropical legume species.

Grasses

Major species of Gramineae which have been adapted for use in sown tropical pastures are listed in the International Rules (ISTA, 1976). They are commonly tested at 20°–35°C alternating temperature using light, with KNO_3 as an alternative treatment for breaking dormancy, although this is of questionable value. Time of final count varies from 14 days (*Chloris gayana* Kunth) to 28 days for most of the other species.

Commercially important species for which official procedures have not yet been listed are set out in table 2.

Unique differences in seedling morphology separate the panicoid and eragrostoid (tropical) from festucoid (temperate) grass seedlings (Silcock, 1980). Tropical species also exhibit definite dormancy characteristics which, despite probable ecological value in the natural state (Mott, 1978), are a considerable problem in quality assessment (Strickland, 1971). Many reports of seed germination responses of tropical grass seeds are found in the literature but field research into the background of crop physiology is of recent origin and has been mainly confined to Kenya and Queensland (Boonman, 1980).

The underlying processes of dormancy in tropical grasses would be as little understood as they are in many other species. Temperature and light are major interactive factors, the operation of which has been reviewed by Toole (1973) as has that of the likely endogenous factors involved (Ketring, 1973).

Useful information on the mechanism of dormancy in tropical grasses has come from cutting and seed coat removal experiments on *Aristida contorta* F. Muell (Mott, 1974) and *Brachiaria ruziziensis* Germain and Evrard (Renard and Capelle, 1976). This work has implicated the seed coat as a competitor with the embryo for oxygen and probably serves to explain the promotive effect of sulphuric acid on germination of a number of species: *Panicum maximum* Jacq. (Smith, 1971), *Brachiaria decumbens* Stapf and *Brachiaria humidicola* (Rendle) Schweicherdt (Atalla and Torello, 1979; C. H. S. Beavis, unpublished data). Other species show little response to sulphuric acid (Cresswell and Nelson, 1972; Harty, 1972). Where sulphuric acid is promotive there are complex interactions with temperature, light and KNO_3 (Smith, 1979). The possible involvement of auxins has been indicated by responses to boron (Cresswell and Nelson, 1972).

The known empirical methods for breaking dormancy were reviewed by Maguire (1975) but it is evident that much additional study is needed to develop effective germination promotion techniques (Maguire, 1976). There are reasons however to believe that this may not be a realistic objective for tropical grasses.

Data on the effect of light are conflicting. Both promotive and inhibitory effects of light have been noted in *Panicum* spp. (Shimizu, 1979). Such effects may be related to different harvesting and drying procedures. Cresswell and Grime (1981) have shown that seeds maturing within green tissues have most of their phytochrome in its active form and thus would be dependent on light to germinate. Exposure to unfiltered light prior to drying would convert phytochrome to the inactive form and remove germinative dependence on light. The response therefore may be strongly affected by factors such as seed maturity at harvest, removal or retention of glumes, and exposure to light during drying, all of which would be virtually impossible to standardise.

Modification of such responses may be possible through recurrent selection (Wright, 1978) but Thompson and Fox (1976) and Thompson (1981) have expressed doubt that plant domestication has ever involved significant modification to germination behaviour. Should such behaviour in tropical grasses be genetically immutable and controlled in its expression by variable environmental influences, then attempts at the total removal of dormancy by means of standardised laboratory procedures are unlikely to be successful.

Viability tests

The recalcitrant nature of post harvest dormancy in these species has highlighted the value of viability testing using tetrazolium. Good agreement between viability estimates and actual germination have been recorded for *Panicum maximum* (Carvalho and Toledo, 1976; Rocha, Marcos Filho and Toledo, 1977; Harty, Hopkinson and English, 1983). Viability measurement has also been found effective for a number of other species (R. L. Harty, unpublished data).

The use of viability testing should be encouraged for official reporting but attention should be given to defining and standardising the expression of viable, germinable and dormant seed (Larsen, 1977).

The Future of germination testing in the tropics

Tropical agriculture has been described as presenting 'an obvious and confusing polyglot of agricultural practices some generally backward, but sometimes also surprisingly good' (Williams, 1975). If seed testing is to be relevant in any agricultural situation its procedures must be compatible with the processes of seed production and distribution.

Different practices are likely to apply to the more uniquely tropical crops such as oil-palm, rubber and tropical fruit species rather than to the more ubiquitous species of 'thermophilic' crops (Gelmond, 1978) such as maize, sorghum and soybeans which are of broad latitudinal significance despite their origins in the tropics. Methods of cropping here will be influenced by the more successful agronomic practices of high energy agriculture and broadly imitative of them.

The development of seed testing in tropical regions will require consideration of both the species of plants which are not being traded at present but which hold promise for wider exploitation, and the agricultural practices involved in their production.

A general problem of germination testing throughout the world is the variation between tests performed in different laboratories. In the tropics the problem is particularly acute because of variation in available controlled environment facilities, inadequate sampling procedure, substrata phytotoxicity or lack of uniform judgement (Verma, Arora and Jethani, 1976). Obviously, here as elsewhere the maintenance and development of seed testing will be dependent upon the level of available government support.

Germination research and the practice of seed testing

The accumulated references to seed germination in world literature are now something in excess of 22,000. Over 800 references to work published between 1970 and 1979 were cited in general literature reviews for this period (Johnston, 1975b, 1977, 1979; Lovato, 1981). Only a small proportion of these would apply directly to seed testing and fewer still would deal with germination testing of uniquely tropical species. On the other hand a wide range of germination research topics impinge on the work of seed analysts and it is useful if not necessary for them to be aware of progress in the understanding of seed behavioural characteristics. The multiplicity of data and observations on germination does not mean that its underlying processes are proportionately better understood (Evenari, 1981), nevertheless seed testing must be ever ready to apply the significant findings of research when these can be readily integrated with the objectives of quality assessment programmes.

References

Abdul-Baki, A. A. and Anderson, J. D. (1973). Vigor determination in soybean seed by multiple criteria. *Crop Science,* **13**, 630–633.

Agrawal, P. K. and Kaur, S. (1975). Standardisation of the tetrazolium test for ragi (*Eleusine coracana*) seeds. *Seed Science and Technology,* **3**, 565–568.

Arora, N. and Banerjee, S. K. (1978). Seed testing procedure for finger millet (*Eleusine coracana*). *Seed Research,* **6**, 158–160.

Atalla, L. M. P. and Tosello, J. (1979). Observations on dormancy in two species of *Brachiaria: Brachiaria decumbens* and *Brachiaria humidicola* under laboratory conditions. *Cientifica,* **7**, 353–355.

Ballard, L. A. T. (1972). High sensitivity to temperature of the germination responses of seeds of Townsville Stylo (*Stylosanthes humilis* H.B.K.). *Proceedings of the International Seed Testing Association*, **37**, 779–791.

Ballard, L. A. T. and Buchwald, T. (1974). Viability test for Townsville Stylo. *Rural Research in CSIRO.*; **77**, 12.

Bavappa, K. V. A. and Jacob, V. J. (1981). A model for mixed cropping. *Ceres - Rome, No. 81*, **14**, 44–46.

Boonman, J. G. (1980). Seed production of tropical grasses in Kenya. In *Seed Production* (ed. P. D. Hebblethwaite) pp. 215–227, Butterworths, London.

Butler, J. E. (1975). Germination of *Stylosanthes humilis* (Townsville Stylo) in short cycles of alternating temperature. *Seed Science and Technology*, **3**, 523–528.

Carvalho, N. M. de and Toledo F. F. de (1976). Determination of the germination of guinea grass (*Panicum maximum* Jacq.) seed by means of tetrazolium 1. Proposal of a working method. *Cientifica*, **4**, 185–190.

Chin, H. F. (1975). Germination and storage of rambutan (*Nephelium lappaceum*) seeds. *Malaysian Agricultural Research*, **4**, 173–180.

Chin, H. F. and Mohd. Lassim, M. B. (1980). Plantation crop seed production in Malaysia. In *Seed Production* (ed. P. B. Hebblethwaite), pp. 449–471, Butterworths, London.

Chua, S. E. and Teoh, T. S. (1973). Propagation by approach grafting and woody cutting of some tropical fruit and ornamental trees. *Singapore Journal Primary Industries*, **1**, 87–95.

Comont, G. and Jacquemard, J. C. (1977). Germination of oil palm seeds (*E. guineensis*) in polythene bags 'dry heat' method. *Oléagineux*, **32**, 151–153.

Cresswell, C. F. and Nelson, H. (1972). The effect of boron on the breaking, and possible control of dormancy of seed of *Themeda triandra* Forsk. *Annals of Botany*, **36**, 771–780.

Cresswell, E. G. and Grime, J. P. (1981). Induction of a light requirement during seed development and its ecological consequences. *Nature, London*, **291**, 583–585.

Cseresnyes, Z. (1979). Studies on the duration of dormancy and methods of determining the germination of dormant seeds of *Helianthus annuus*. *Seed Science and Technology*, **7**, 179–188.

Das, N. K. and Baruah, H. K. (1974). Physiology of areca nut (*Areca catechu* L.). *Journal of Plantation Crops*, **2**, 7–13.

Duke, J. A. (1981). *Handbook of legumes of world economic importance*. Plenum Press, New York, London.

Ellis, R. H. and Roberts, E. H. (1980). Towards a rational base for testing seed quality. In *Seed Production* (ed. P. B. Hebblethwaite), pp. 605–635, Butterworths, London.

Ellis, R. H. and Roberts, E. H. (1981). Germination of stored cassava seed at constant and alternating temperature. *Annals of Botany*, **44**, 677–684.

Evenari, M. (1981). The history of germination research and the lesson it contains for today. *Israel Journal of Botany*, **29**, 4–21.

Febles, G. and Padilla, C. (1971). Effect of sulphuric acid on the germination and early growth of *Glycine wightii*. *Cuban Journal of Agricultural Science*, **11**, 105–112.

Gaber, S. D., Abdalla, F. H. and Mahdy, M. T. (1974). Treatments affecting dormancy in sweet sorghum seed. *Seed Science and Technology*, **2**, 305–316.

Gelmond, H. (1978). Physiological aspects of seed germination. *Seed Science and Technology*, **6**, 625–639.

Gelmond, H. (1979). A review of factors affecting seed quality distinctive to cotton seed production. *Seed Science and Technology*, **7**, 39–46.

Goldbach, H. (1979). Germination and storage of *Bixa orellana* seeds. *Seed Science and Technology*, **7**, 399–402.

Goldbach, H. and Aguilera Vizcarra, H. (1980). Some observations on tetrazolium testing of coffee seed (*Coffea arabica* and *C. canephora*). *Turrialba*, **30**, 223–226.

Gribbin, J. (1978). World agriculture: the need for a scientific base. *New Scientist*, **80**, 514.

Harper, J. L. (1977). *Population biology of plants*. Academic Press London, New York and San Francisco.

Hartley, W. (1979). *A Checklist of Economic Plants in Australia*. CSIRO, Melbourne.

Harty, R. L. (1972). Germination requirements and dormancy effects in seed of *Urochloa mosambicensis*. *Tropical Grasslands*, **6**, 17–24.

Harty, R. L., Hopkinson, J. M. and English, B. H. (1983). Germination, dormancy and longevity in stored seed of *Panicum maximum*. *Seed Science and Technology*, (in press).

Heit, C., Robinson, R. W. and Mishanec, W. (1978). Dormancy of *Cucumis* species. *Cucurbit Genetics Co-operative*, (New York Agricultural Experiment Station, Geneva, USA **1**, 36–37.

Hopper, N. W. and Hinton, H. R. (1980). The use of electrical conductivity as a measure of cottonseed quality. *Agronomy Abstracts, 72nd Annual Meeting American Society of Agronomy*.

Hsieh, S. U. (1973). The test on the dormancy of rice varieties. *Taiwan Agricultural Journal*, **9**, 181–186.

Humphreys, L. R. (1979). *Tropical pasture seed production*. FAO, Rome.

Humphreys, L. R. (1980). Deficiencies of adaptation of pasture legumes. *Tropical Grasslands*, **14,** 153–158.

International Seed Testing Association (1976). International rules for seed testing. Rules, Annexes 1976. *Seed Science and Technology*, **4**, 3–177.

Johnston, M. E. H. (1972). Report of the working group for the germination of tropical and sub-tropical seeds. *Proceedings of the International Seed Testing Association*, **37**, 355–359.

Johnston, M. E. H. (1975a). Report of the working group on the germination of tropical and sub-tropical seeds. *Seed Science and Technology*, **3**, 94–99.

Johnston, M. E. H. (1975b). Seed germination. *Advances in Research and Technology of Seeds*, **1**, 8–33.

Johnston, M. E. H. (1977). Germination of seed. *Advances in Research and Technology of Seeds*, **3**, 7–53.

Johnston, M. E. H. (1978). Report of the germination committee working group on tropical and sub-tropical seeds 1974–1977. *Seed Science and Technology*, **6**, 193–197.

Johnston, M. E. H. (1979). Germination of seed. *Advances in Research and Technology of Seeds*, **4**, 43–48.

Johnston, M. E. H. and Harty, R. L. (1981). Report of the germination committee working group on tropical and sub-tropical seeds 1977–1980. *Seed Science and Technology*, **9**, 137–140.

Jones, R. M. and Evans, T. R. (1977). Soil seed levels of *Lotononis bainesii, Desmodium intortum* and *Trifolium repens* in sub-tropical pastures. *Journal of the Australian Institute of Agricultural Science*, **43**, 164–166.

Joshi, R. K., Mishra, S. D. and Gaur, B. K. (1978). Release of dormancy in Trombay groundnut seeds by plant growth regulators. *Acta Botanica Indica* **6**, (Supplement) 7–15.

Kamil, J. (1977). Relation of specific gravity of rice (*Oryza sativa* L.) seed to viability and vigour. *Panitia Koordinasi penelitian pertanian*, **7**, 13.

Ketring, D. L. (1973). Germination inhibitors. *Seed Science and Technology*, **1**, 305–324.

Ketring, D. L. and Morgan, P. W. (1969). Ethylene as a component of the emanations from germinating peanut seeds and its effect on dormant virginia-type seeds. *Plant Physiology, Lancaster*, **44**, 326–330.

Kirpal Singh, K., Chadha, K. L., and Gupta, M. R. (1973). *Ber cultivation*. Punjab Agricultural University, Ludhiana.

Kordan, H. A. (1976). Adventitious root initiation and growth in relation to oxygen supply in germinating rice seedlings. *New Phytologist*, **76**, 81–86.

Kordan, H. A. (1977). Coleoptile emergence in rice seedlings in different oxygen environments. *Annals of Botany*, **41**, 1205–1209.

Kulkarni, G. N. and Basavaraju, V. (1976). Firm seeds in the finger millet (*Eleusine coracana*, G.). *Current Science*, **45**, 425.

Larsen, A. L. (1977). Determining and reporting of dormant seed. *Newsletter Association of Official Seed Analysts North America*. **51**, 33–34.

Lassim Mohd., M. B. and Chin, H. F. (1980). Rice seed production in west Malaysia. In *Seed Production* (ed. P. D. Hebblethwaite), pp. 357–369, Butterworths, London, Boston.

Lovato, A. (1981). Germination of seeds. *Advances in Research and Technology of Seeds*, **6**, 86–120.

Lugo, A. E. and Brown, S. (1981). Tropical lands: popular misconceptions. *Mazingira*, **5**, 10–19.

MacKay, D. B. (1978). The laboratory germination test. *Acta Horticulturae*, **83**, 97–102.

McDonald, M. B. and Wilson, D. O. (1980). ASA-610 ability to detect changes in soybean seed quality. *Journal of Seed Technology*, **5**, 56–66.

McIvor, J. G. (1976). Germination characteristics of seven *Stylosanthes* species *Australian Journal of Experimental Agriculture and Animal Husbandry*, **16**, 723–728.

Maguire, J. D. (1975). Seed dormancy. *Advances in Research and Technology of Seeds*, **1**, 44–53.

Maguire, J. D. (1976). Seed dormancy. *Advances in Research and Technology of Seeds*, **2**, 9–26.

Malo, S. E. and Martin, F. W. (1979). Cultivation of neglected tropical fruits with promise. Part 7. The durian. *United States Department of Agriculture, Agricultural Research Service ARS-SUS*.

Marx, J. L. (1977). Amaranth: a comeback for the food of the Aztecs? *Science, New York*, **198**, 40.

Matthews, S. (1980). Controlled deterioration; a new vigour test for crop seeds. In *Seed Production* (ed. P. D. Hebblethwaite), pp. 647–660. Butterworths, London.

Meadly, G. R. W. (1972). Testing tropical and sub-tropical seed species – introduction. *Proceedings of the International Seed Testing Association*, **37**, 713–718.

Meighan, C. W., Pendergast, D. M., Swartz, B. K. (Jr.) and Wissler, M. D. (1958). Ecological interpretation in archaeology: Part II. *American Antiquity*, **XXIV**, 131–147.

Mok, C. K. (1972a). Sample size for moisture and viability testing of oil palm (*Elaeis guineensis* Jacq). *Proceedings of the International Seed Testing Association*, **37**, 751–761.

Mok, C. K. (1972b). The tetrazolium test for evaluating the viability of oil palm (*Elaeis guineensis* Jacq). seed. *Proceedings of the International Seed Testing Association*, **37**, 771–778.

Morton, J. F. (1966). The soursop, or guanabana (*Annona muricata* Linn.). *Proceedings of the Florida State Horticultural Society*, **79**, 355–366.

Mott, J. J. (1974). Mechanisms controlling dormancy in the arid zone grass *Aristida contorta*. I. Physiology and mechanisms of dormancy. *Australian Journal of Botany*, **22**, 635–645.

Mott, J. J. (1978). Dormancy and germination in five native grass species from savannah woodland communities of the Northern Territory. *Australian Journal of Botany*, **26**, 621–631.

Mott, J. J. and McKeon, G. M. (1979). Effect of heat treatments in breaking hardseededness in four species of *Stylosanthes*. *Seed Science and Technology*, **7**, 15–25.

Nangju, D., Wien, H. C. and Ndimande, B. (1980). Improved practices for soya bean seed production in the tropics. In *Seed Production* (ed. P. D. Hebblethwaite), pp. 427–448. Butterworths, London.

National Academy of Sciences (1974). *Food science in developing countries: a selection of unsolved problems* (ed. E. F. Pariser). National Academy of Sciences, Washington D. C..

National Academy of Sciences (1975). *Underexploited tropical plants with promising economic value*. National Academy of Sciences, Washington D. C.

National Academy of Sciences (1977). Leucaena: *promising forage and tree crop for the tropics*. National Academy of Sciences, Washington D. C..

National Academy of Sciences (1979). *Tropical legumes: resources for the future*. National Academy of Sciences, Washington D.C..

Noiret, J. M. (1968). Germination industrielle des graines de palmier à huile. (Large scale germination of oil palm). *Oléagineux*, **23**, 293–298.

Odetola, A. (1974). Heat requirement of oil palm seeds for germination. *Journal of the Nigerian Institute for Oil Palm Research*, **5**, 79–84.

Okusanya, O. T. (1979). Quantitative analysis of the effects of photoperiod temperature, salinity and soil types on the germination and growth of *Corchorus olitorius*. *Oikos*, **33**, 444–450.

Omran, A. F., El-Bakry, A. M. and Gawish, R. G. (1980). Effect of soaking seeds in some growth regulator solutions on the growth, chemical constituents and yield of okra. *Seed Science and Technology*, **8**, 161–168.

Pe, W., Hill, M. J. and Johnston, M. E. H. (1975). Effect of seed storage and seed treatment on the germination of *Centrosema pubescens* (*Centro*) seeds. I Acid treatment and mechanical scarification. *New Zealand Journal of Experimental Agriculture*, **3**, 81–89.

Phipps, R. H. (1973). Methods of increasing the germination percentage of some tropical legumes. *Tropical Agriculture in Trinidad*, **50**, 291–296.

Renard, C. and Capelle, P. (1976). Seed germination in ruzizi grass (*Brachiaria ruziziensis* Germain, Evrard). *Australian Journal of Botany*, **24**, 437–446.

Roberts, E. H. and Ellis, R. H. (1980). Seed physiology and seed quality in soya bean. In *Advances in Legume Science* (eds R. J. Summerfield and A. H. Bunting), pp. 297–311. Kew.

Rocha, V. A. DA; Marcos Filho, J.; Toledo, F. F. de (1977). Measurement of seed viability in guinea grass by the tetrazolium method. *Revista de Agricultura, Piracicaba Brazil*, **52**, 19–23.

Ruberte, R. M. (1977). Species suitable for the production of edible seed sprouts. *Journal of Agriculture University of Puerto Rico*, **61**, 262–264.

Scott, C. L. (1980). Research seeds and concerns as viewed by farmers. In *Proceedings, Beltwide Production-Mechanisation Conference,* The Cotton Foundation, Memphis, Tennessee.

Sento, T. (1976). Studies on the germination of palm seeds. *Memoirs of the College of Agriculture, Ehime University,* **21**, 1–78.

Sharir, A. (1978). Some factors affecting dormancy breaking in peanut seeds. *Seed Science and Technology,* **6**, 655–660.

Shimizu, N. (1979). Studies on dormancy and germination of seeds in grasses of *Panicum* species. 1. Light-temperature response in germination and the dormancy breaking effect of metabolic inhibitors. *Bulletin of the National Grassland Research Institute,* **14**, 94–101.

Silcock, R. G. (1980). Seedling characteristics of tropical pasture species and their implications for ease of establishment. *Tropical Grasslands,* **14**, 174–180.

Singh, A., Karna I. and Verma, K. (1972). Germination methods for jute seed. *Proceedings of the International Seed Testing Association,* **37**, 793–796.

Singh, C., Bhan, A. K. and Kaul, B. L. (1974). The role of some physical and chemical agents in the improvement of seed germination in *Datura innoxia. Seed Science and Technology,* **3**, 421–426.

Singh, S. M. and Kaisuwan, T. (1971). Effect of interval between extraction and sowing, and container for storage on the germination of seed stones and growth of seedlings of mango (*Mangifera indica* Linn.). *Tropical Abstracts,* **28**, 272–273.

Smith, A. M. and Cook, R. J. (1974). Implications of ethylene production by bacteria for biological balance of soil. *Nature, London,* **252**, 703–705.

Smith, C. J. (1971). Seed dormancy in sabi panicum. *Proceedings of the International Seed Testing Association,* **36**, 81–97.

Smith, R. L. (1979). Seed dormancy in *Panicum maximum* Jacq. *Tropical Agriculture, in Trinidad,* **56**, 233–239.

Soepadmo, E. and Eow, B. K. (1977). The reproductive biology of *Durio zibethinus* Murr. *Gardens Bulletin, Singapore,* **29**, 25–33.

Souza, F. H. D. de and Marcos Filho, J. (1975). A comparison of methods for assessing the vigour of seeds of sorghum (*S. vulgare* Pers.) *Anais das Escola Superior de Agricultura Luiz de Queiroz,* **32**, 369–383.

Spain, G. E. (1976). Seed vigor, a seed control official's viewpoint. *Journal of Seed Technology,* **1**, 13–17.

Strickland, R. W. (1971). Seed production and testing problems in tropical and sub-tropical pasture species. *Proceedings of the International Seed Testing Association,* **36**, 189–199.

Tao, K-L. J. (1978). An evaluation of alternative methods of accelerating ageing seed vigour test for soybeans. *Journal of Seed Technology,* **3**, 30–40.

TeKrony, D. M. and Egli, D. B. (1977). Relationship between laboratory indices of soybean seed vigour and field emergence. *Crop Science,* **17**, 573–577.

TeKrony, D. M., Egli, D. B. and Balles, J. (1980). The effect of the field production environment on soya bean seed quality. In *Seed Production* (ed. P. D. Hebblethwaite), pp. 403–425, Butterworths, London.

Teng, Y. T. (1980). Studies on the storage of some Malaysian tropical fruit seeds. *Seminar Nasional Buah-Buahan Malaysia Universiti Pertanian Malaysia Paper no. 10.*

Thompson, P. A. (1981). Ecological aspects of seed germination. *Advances in Research and Technology of Seeds,* **6**, 9–42.

Thompson, P. A. and Fox, D. J. C. (1976). The germination responses of vegetable seeds in relation to their history of cultivation by man. *Scientia Horticulturae,* **4**, 1–14.

Toole, V. K. (1973). Effects of light, temperature and their interactions on the germination of seeds. *Seed Science and Technology,* **1**, 339–396.

Turner, J. H. and Ferguson, D. (1972). Field performance of cotton grown from filled and partially filled seeds. *Crop Science,* **12**, 868–871.

Tyagi, M. C., Singh, M. P. and Bammi, R. K. (1973). The effect of temperature on seed germination of *Dioscorea* species. *Planta medica,* **24**, 294–296.

Valio, L. F. M. (1976). Germination of coffee seeds (*Coffea arabica* L. cv. Mundo Nov. *Journal of Experimental Botany,* **27**, 983–991.

Verma, M. M. and Arora, N. (1978). Further studies on seed testing procedures for jute (*Corchorus capsularis* and *C. olitorius*) seeds. *Seed Research,* **6**, 151–157.

Verma, M. M., Arora, N. and Jethani, I. (1976). Problems in maintaining uniformity in seed testing results, germination. *Seed Research,* **4**, 94–99.

Whitman, W. F. (1980). Growing and fruiting the langsat in Florida. *Proceedings of the Florida State Horticultural Society,* **93**, 136–140.

Williams, C. (1975). *The Agronomy of major tropical crops.* OUP, Kuala Lumpur and New York.

Wright, L. N. (1978). Recurrent selection for changing gene frequency of germination rate in blue panic grass. *Crop Science,* **18**, 789–791.

Yasue, T. (1973). Effect of moisture content of seeds and soaking in water on breaking dormancy in indica rice. *Research Bulletin of the Faculty of Agriculture Gifu University,* **34**, 1–10.

Yasue, T. and Asia, Y. (1970). The influence of storage conditions on dormancy and germination in rice seed. *Research Bulletin of the Faculty of Agriculture Gifu University,* **29**, 67–69.

Yaycock, J. Y., Jan, H. and Vanderlip, R. L. (1975). Temperature, time and NH_4 Cl concentration in vigor testing of sorghum seed. *Agronomy Journal,* **67**, 241–242.

O'Dowd T. and Dobie P. (1983), *Seed Sci. & Technol.*, **11**, 57–75

Reducing viability losses in open seed stores in tropical climates

T. O'DOWD and P. DOBIE

Tropical Products Institute, Storage Department, (Tropical Stored Products Centre), London Road, Slough, Berks SL3 7HL, UK

(Accepted October 1982)

Summary

Ways in which viability losses can be reduced in open or naturally ventilated seed stores are described. The correct physical conditions for seed storage can be predicted enabling the optimum combination of building design and drying facility to be chosen. The conditions prevailing in an open store may favour the development of serious infestations of insect pests. Methods of protecting seed from insect attack are discussed.

Résumé

Réduction des pertes de viabilité dans les entrepôts ouverts de semences sous des climats tropicaux

On décrit les façons dont on peut réduire les pertes de viabilité dans les entrepôts de semences ouverts ou naturellement ventilés. Les conditions physiques correctes pour la conservation des semences peuvent être prévues qui permettent le choix de la combinaison optimum du plan du bâtiment et de l'installation de séchage. Les conditions régnant dans un entrepôt ouvert peuvent favoriser le développement de graves infestations d'insectes parasites. On discute les méthodes de protection des semences contre les attaques des insectes.

Zusammenfassung

Zur Reduzierung von Einbußen der Lebensfähigkeit bei offener Saatgutlagerung in tropischen Klimaten

Es wird die Art und Weise beschrieben, mit welcher Verluste an Lebensfähigkeit bei offener oder natürlich belüfteter Saatgutlagerung reduziert werden können. Die richtigen physikalischen Bedingungen für Saatgutlagerung können vorhergesagt werden, welche die optimale Kombination der Gebäudekonstruktion und der zu wählenden Trocknungseinrichtungen ermöglichen. Die Verhältnisse, welche in einem offenen Lager vorherrschen, können die Entwicklung schweren Befalls durch Schadinsekten begünstigen. Methoden zum Schutz des Saatguts vor Insektenbefall werden diskutiert.

Introduction

This paper is concerned with the reduction of seed viability losses in tropical climates during storage in conventional buildings (where viability refers to the capability of a seed to develop into an acceptable seedling). These buildings are naturally ventilated and are known as open stores (as opposed to conditioned atmosphere stores in which the internal climate is controlled using air-conditioning plant).

The main causes of viability losses in open stores, identified by Delouche, Matthes, Dougherty and Boyd (1973) are (a) low initial viability of seed, (b) incorrect conditions during storage and (c) insect attack.

Our objective is to consider ways in which the correct conditions in open seed storage can be achieved, taking initial viability into account, and also how viability losses caused by insect attack can be avoided or reduced.

2. The necessary physical conditions

Suitable combinations of seed temperature and moisture content are required to maintain acceptable viability levels during open storage. Delouche *et al.* (1973) have provided broad guidelines for safe storage in terms of store temperature, relative humidity and seed moisture content for a wide range of seed types in the tropics and sub-tropics. These data are useful because they are based on field experience and cover many conditions. The data for short term and intermediate storage are reproduced in table 1. Although the authors stress the importance of initial viability, their data make no specific allowance for the effect on safe storage conditions of the often low initial viability of seed received into store in the tropics and sub-tropics.

Table 1. Safe conditions for storage for good quality seed.

		Maximum seed moisture content	
Period	Conditions	Cereals	Oil seeds
A. Short Term (1–9 months)	(a) 30°C–50% RH	12%	8%
	(b) 20°C–60% RH	13%	9.5%
	(c) Other combinations as favourable as (a) and (b)		
B. Intermediate (18 months)	(a) 30°C–40% RH	10%	7.5%
	(b) 20°C–50% RH	12%	8%
	(c) 10°C–60% RH	12%	9%
	(d) Other combinations as favourable as (a) and (b)		

Source: Delouche, Matthes, Dougherty and Boyd (1973)

The seed viability equations of Ellis and Roberts (1980a) allow for the effect of different levels of initial viability on the necessary physical storage conditions, although their data cover fewer seed types. If a given level of initial and final viability is chosen for a given seed type, the equations can be expressed simply in the form of a graph (figure 1). In the example given, which is for maize (*Zea mays* L.) seed,

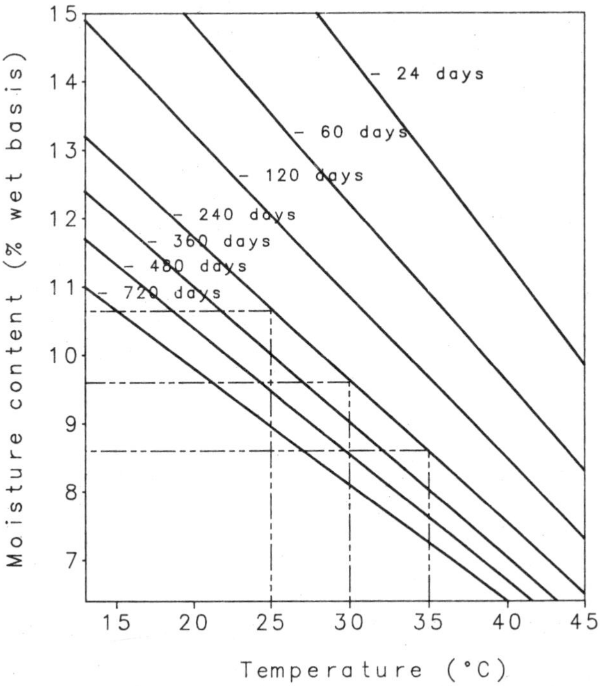

Figure 1. Seed life in days for maize. (Time taken for viability to drop from 95–85%. Combined effects of moisture content and temperature). Source: Ellis, personal communication (1981)

the initial viability is 95% and the final viability is 85%. The final viability may appear high but it is the legal requirement in several countries overseas. The necessary conditions of temperature and moisture content are shown for storage lives of 24 to 720 days. Thus to achieve 240 days safe storage any number of temperature and moisture content combinations are possible: 8.5% moisture content at 35 °C, 10.7% at 25 °C, etc. As temperature falls, the recommended moisture content rises. At 30 °C a moisture content of 9.6% is recommended, while Delouche *et al.* recommend a *maximum* moisture content of 12% at this temperature. The recommendations of Ellis and Roberts are more stringent because they take account of initial viabilities which are an indication of seed quality which affects the subsequent period of seed survival and will be used in this paper for this reason. For the limited range in conditions where viability falls from 95–85% and storage temperatures of 20 °C to 45 °C obtain, the relationship between storage life (S_L), moisture content (wet basis) (m), effective seed temperature (T_E) for maize can be expressed in a simplified equation (Ellis, 1981):

$$S_L = \text{antilog}_{10}\,(6.49 - 0.269\text{m} - 0.052\ T_E) \qquad (1)$$

Equation (1) shows that if the effective temperature is known the required moisture content can be found, for a given seed life (the effective temperature takes account of the swings in temperature that occur in open storage and is discussed below). It is necessary to predict the effective temperature in an open store in order to arrive at the correct seed moisture content before the store is built, so that the feasibility of drying to this level can be considered.

3. Prediction of effective seed temperature

It has been shown that 'safe' storage conditions for seed consist of a combination of seed temperature and moisture content. Seed temperature is a function of climate and building design, so once the open seed store has been built the seed temperature cannot be controlled to any degree and is fixed. This in turn determines the moisture content required for safe storage.

If the seed temperature can be predicted at the design stage the safe level of moisture content can also be predicted and the feasibility of drying seed to that level examined. In many cases open stores have been built with an inadequate level of insulation which results in seed temperatures that are so high that unattainable drying levels would be necessary to achieve safe storage.

The following method of prediction is suggested using standard heat flow equations (Petherbridge, 1974) modified to allow for the effect of grain in the building. With no solar radiation on the building and a constant rate of ventilation the mean seed temperature (t_g), the mean outside temperature over 24 hours (t_{ao}) and the mean inside temperature (t_{ei}) would all be equal:

$$t_g = t_{ei} = t_{ao} \tag{2}$$

With constant solar radiation the mean grain temperature (t_g) would be influenced by the mean outside temperature and the effects of solar radiation, conduction and ventilation on heat flow:

$$t_g = t_{ei} = t_{ao} + \frac{\Sigma AUR_{so}\,(aI_t - eI_L)}{AU + C_v} \tag{3}$$

where for each surface of the building

A = area (m^2)
U = transmittance ($Wm^{-2}\,°C$)
R_{so} = surface resistance ($m^2\,°C\,W^{-1}$)
a = absorption coefficient
I_t = solar radiation intensity (Wm^{-2})
e = emittance
I_L = Long wave radiation (Wm^{-2})
C_v = ventilation conductance ($W\,°C^{-1}$)

The equation shows that the mean seed temperature (t_g) will always be higher than the mean outside temperature (t_{ao}) by an amount which is influenced by several factors. The numerator contains the solar radiation term I_t and the effect of this term is modified by multiplying it by the solar absorption coefficient (a), the surface resistance (R_{so}), the transmittance (U) and the area (A). The designer can influence the choice of materials with low absorption coefficients and low transmittances, while there is less choice over surface resistance, which is largely controlled by the exposure of the site, or the surface area which can only be influenced marginally by changing the orientation of the building. The denominator contains a conductive loss term (ΣAU) and a ventilation term (C_v). This implies that the higher the rates of conduction and ventilation the lower the heat gain will be, but when the effect of diurnal swings in solar radiation is considered, this conclusion is modified:

$$\tilde{t}_g = \tilde{t}_{ei} = \frac{(\Sigma fAU + C_v)\,\tilde{t}_{ao} + \Sigma fAUR_{so}\,(aI^1{}_t - eI_L)}{\Sigma AY + C_v + \frac{kK}{2w}} \qquad (4)$$

$\tilde{t}_{ao}$ = swing in outside temperature (°C)
$\tilde{t}_g$ = swing in grain temperature (°C)
$\tilde{t}_{ei}$ = swing in internal temperature (°C)
f = decrement factor for each surface
Y = admittance factor for each surface (Wm^{-2} °C)
K = weight of seed affected by swing in temperature (kg)
k = specific heat of grain (J $°C^{-1}$ kg^{-1})
w = time lag assumed for grain to reach swing temperature (s)
$I^1{}_t$ = radiation swing for each surface (Wm^{-2})

It can be seen that in equation (4) the term C_v appears in the denominator and the numerator. The implication here is that increased ventilation during the warmest parts of the day could cause bigger rather than smaller temperature swings. In addition, equation (4) introduces new parameters in order to predict the swing in grain temperatures. The decrement factor f is present in both terms in the numerator, and is the damping effect of the thermal capacity of the roof and walls. Basically the thicker the walls or roof the greater is the damping effect and the smaller the swing in temperature. In the denominator the term ΣAY is introduced. Y is the admittance of the materials used to construct the store, and is a measure of the ability of the building component to store and release energy (heat) over the daily cycle. Brickwork and concrete have high admittances while lightweight materials have low admittances. Thus in addition to choosing materials with low transmittances (U values) and low absorbtivities (a values) the designer must also consider decrement values and admittances to achieve low swings in temperature ($\tilde{t}_g$) as well as low mean temperature (t_g). Small swings in seed temperatures are important because the high temperatures have a disproportionately large effect on the 'effective' seed temperature used in equation (1) (Roberts, 1961).

This is shown in equation (5) below:

$$T_E = \log [\{\text{antilog}_{10} (C_H\ t_g \max) + \text{antilog}_{10} (C_H\ t_g \min)\}/2]/C_H \quad (5)$$

where

T_E = effective seed temperature

C_H = constant, 0.06 for *Zea mays*

The thermal effect of the seed in store on the swings in temperature is allowed for in equation (4) by including the affected seed in the heat balance equation. The term $\frac{kK}{2w}$ represents the heat flow (in $W°C^{-1}$) needed to cause the swing in temperature in the affected seed. It is assumed that the temperature swing will only affect the outer layer of seed in the stacks of bags. The indications are that this layer will be about 25 cm deep and the time lag approximately 6–10 hours.

The effective temperature so calculated strictly applies only to the affected seed on the outside of the stacks but is used here for the whole seed stock. This simplication is made because there is no easy way of distinguishing 'inside' from 'outside' bags and legal requirements apply to both. Also stack sizes in seed stores are commonly small so that a high percentage of seed is affected by temperature swings.

The equations (3) and (4) enable the effective seed temperature (T_E) to be predicted. The data required include climatic data and building data, a typical sample of which is shown in appendix 1. In an equatorial humid climate the temperatures predicted for the four different levels of insulation, absortivity, decrement and admittance shown are as follows:

Level of insulation	Seed temperatures °C		
(see appendix 1)	Maximum	Minimum	Effective temperature
R1 W1	47	29	42
R2 W2	38	31	35
R3 W3	34	31	33
R3 W4	32	30	31

The method developed by the UK Building Research Establishment (BRE) for predicting internal temperatures in buildings is well tried. The adapted method shown above for predicting seed temperatures will be tested in the field but the indications are that it can be a useful predictive tool. In the example shown above the effective temperature of seed in a poorly insulated building can be reduced by 11 °C in a well insulated building. The important point is that for a given climate and known building design it is possible to predict what the likely effective seed temperature will be.

4. Prediction of safe seed moisture content

A method for predicting the effective seed temperature in an open store has been described. Once this temperature is known the corresponding moisture content necessary to achieve a given seed life can also be predicted using the seed viability equations or the simplified graphical method referred to earlier. For a limited range of conditions, equation (1) is applicable.

$$S_L = \text{antilog}_{10}\,(6.49 - 0.269m - 0.052\,T_E) \qquad (1)$$

where

S_L = storage life in days
m = moisture content % wet basis
T_E = effective temperature °C

In the example quoted above four different combinations of building materials were employed to predict four different effective temperatures. Using equation (1) it is possible to predict the safe moisture content for maize for a given seed life; in this example a seed life of 150 days is taken:

Level of insulation (see appendix 1)	(Predicted) Effective temperature °C	(Predicted) Safe moisture content % wet basis
R1 W1	42	7.8
R2 W2	35	9.2
R3 W3	33	9.7
R4 W4	31	10.0

The safe moisture contents shown above are all lower than the moisture content which is frequently used in practice, wrongly based on the ambient temperature. These corrected moisture contents take account of the higher than ambient temperatures inevitable in (storage) buildings and the effect of temperature swings. Many of the losses in viability experienced in open stores can be explained by this misconception. However, as shown in the table above, a well insulated building enables seed to be stored safely at a considerably higher moisture content than a poorly insulated building.

The prediction of the safe moisture content for different types of building is essential when planning and designing open storage seed operations. The safe moisture content determines how much drying is required and also what type of packaging is needed, for particular intake and storage conditions.

5. Prediction of amount of drying and packaging needed

The necessary drying capability can be calculated if the safe moisture content is known for a given intake moisture content and known ambient conditions. It is beyond the scope of this paper to describe all of the factors which influence the drying process but three factors are particularly important in seed drying. The first is that to maintain seed viability during drying the seed temperature should not be allowed to rise above about 40 °C and this limits the rate of drying possible with any machine especially when ambient relative humidity and temperature are high. Secondly, the rate of drying decreases sharply as the required level of moisture content falls. It is much harder to reduce seed from 11% to 9% than from 13% to 11%. Lastly the ambient relative humidity has a sharp effect on drying capacity: where high relative humidities obtain, drying below a given moisture content may be impossible and is certainly very expensive. These three factors combine to limit the drying operations which are feasible for seed. For the example quoted above, where there is an intake of 1700 tonnes of grain at 13% moisture content and 500 hours of drying time available the numbers of 1 tonne/hour (nominal) tray driers is estimated as follows:

Level of insulation	Safe moisture content	Number of driers required
R1 W1	7.8	12
R2 W2	9.2	8
R3 W3	9.7	7
R4 W4	10.0	6

With the poorly insulated building (R1 W1) it is likely that the low moisture content required is either technically unobtainable or prohibitively expensive. (It is estimated that the total drying costs per tonne will be £21 at 1980 prices). With the well insulated building (R4 W4) it is likely that the level of moisture content can be achieved (drying costs are estimated at £10/tonne). These considerations are vital to the success of an open storage operation.

The prediction of safe seed moisture content will indicate what type of packaging is required. Over the storage season the store relative humidity may rise above the equilibrium relative humidity of the stored seed, and if so moisture vapour proof packaging will be necessary to prevent the seed absorbing moisture and losing viability. The lower the safe seed moisture content required the more likely the need for special packaging and only if the safe moisture content can be predicted can this element of the planning process be considered.

6. Prediction of costs

Prediction of the effective seed temperatures requires that the thermal properties of the building materials used are known and it is therefore possible to obtain costs of these. Likewise prediction of the drying capacity required enables the costs of drying to be estimated. These two costs combined give the total costs of drying and storage to be estimated for each storage system in this illustrative example:

Storage system	Annual costs per tonne			
Level of insulation	Number of driers	Building insulation	Drying	Total
R1 W1	12	£1.0	£21.0	£22.0
R2 W2	8	£1.5	£14.0	£15.5
R3 W3	7	£2.8	£12.0	£14.8
R4 W4	6	£3.9	£10.0	£13.9

The figures shown above are based on estimates (see appendix 2) and should not be regarded as definitive. Moreover it is unlikely that large numbers of low capacity driers would be considered for these operations (smaller numbers of large capacity driers would be used if suitable alternatives could be found). However the figures illustrate that it is possible to predict costs and so arrive at the optimum combination of building insulation and drying capacity to achieve the required storage life at least cost.

7. A mathematical model for prediction

At the Tropical Products Institute (TPI) a mathematical model is being constructed, the object of which is to make possible the attainment of a desired seed life under different intake and climatic environments by predicting the necessary storage conditions. The model is made up of a data bank and a series of information components generated by mathematical equations. A computer is used for the calculations. Figure 2 shows the model's operation, each component represented by a box and each process by an arrow. Effective seed temperatures, safe moisture contents, hours of drying required and costs are generated using information from the data bank.

The advantages of the computer model are that a wide range of seed types, building materials and climatic conditions can be considered simultaneously. In addition the effect of varying one particular parameter can be found easily. This is useful to the designer when there is doubt about the value of one parameter, because the effect of varying a value can be easily shown. For example, the effect of dust (or its absence) on cladding materials will affect seed temperatures and the computer model can predict how much and thus provide a guide to design.

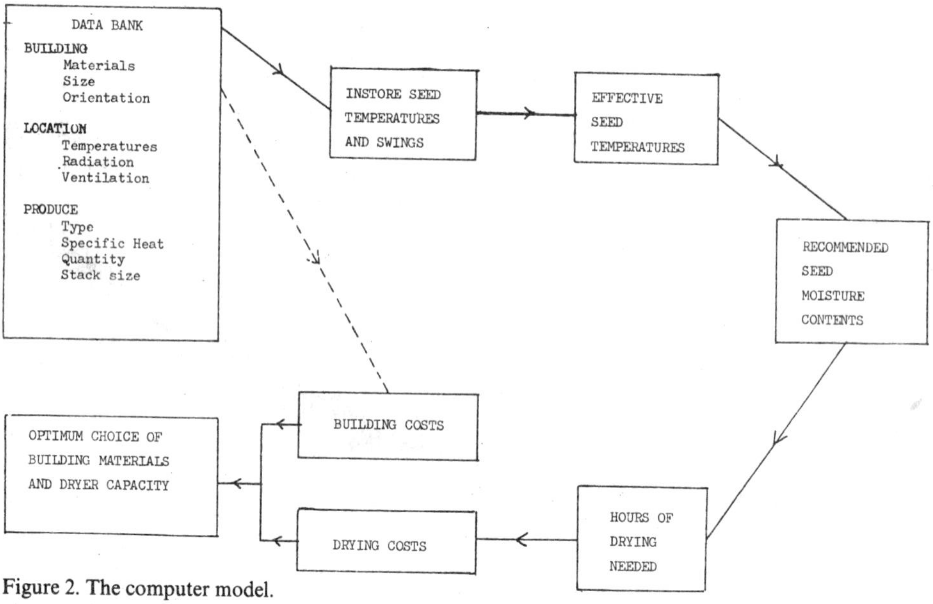

Figure 2. The computer model.

At present the model provides for an arid and a humid tropical climate and a limited range of building materials. The predictions suggest that effective seed temperatures in poorly insulated open stores are well above ambient mean temperatures and that drying requirements are very exacting. In well insulated stores, however, effective seed temperatures approach ambient mean temperatures and drying requirements are technically feasible and more financially acceptable. These indicative results are relevant to the practical problems of seed storage in the tropics. The next step is to verify these results in the field and then enlarge the capacity of the model. If this can be done, and the indications are that it can, the model should provide a useful predictive tool for seed store design.

8. Pests and their control in seed stores

Insect pests can cause considerable damage to stored seeds. They reduce viability by feeding upon the storage tissue, cotyledons and embryos of the seed and by damaging seed coats and permitting the entry of pathogens. If insect numbers are high, their metabolism can cause a considerable increase in temperature, which may become high enough to kill the seed. Their metabolism also increases the moisture content of seed, and their metabolic heat can cause moisture migration within the bulk. Both effects can cause severe moulding and consequential loss of viability. Many species of insect can attack stored seeds (Howe, 1973), but most of them can do so efficiently only if the seed is damaged, damp or of poor quality. The remaining species are important pests of dry, undamaged, high quality seed and are extremely well adapted for breeding

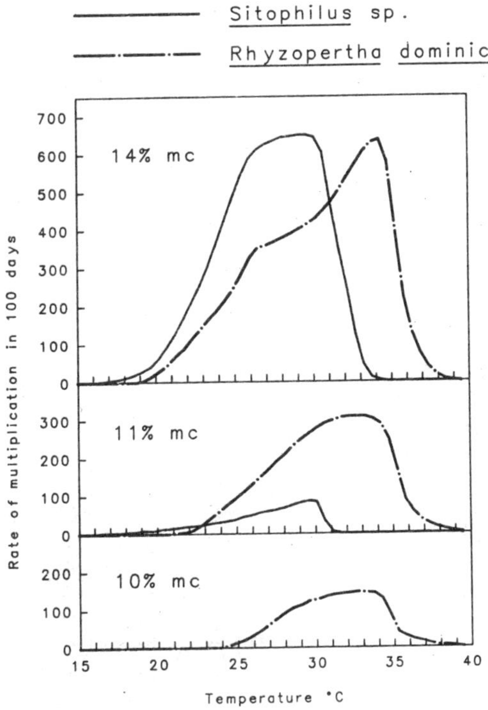

Figure 3. The effects of moisture and temperature upon the multiplication of *Sitophilus oryzae* and *Rhyzopertha dominica*. Source: Birch (1945).

and multiplying in stored seed. They are coloniser species, which may be introduced into seed in small numbers, but, if conditions are suitable, will reproduce at a very great rate and quickly infest an entire consignment. Their development periods under favourable conditions are very short (typically five weeks or less) and many species can reproduce over one hundred times in a single five week generation. Thus it is necessary to store seed under conditions that severely reduce the reproductive potential of pests or to rely upon chemicals to completely eliminate them.

The reproductive potential of pests is limited by the temperature and moisture content of the seed. Figure 3 shows the effect of different temperatures and grain moisture contents upon the multiplication rates of two important pests of cereal seeds, *Sitophilus oryzae* L. and *Rhyzopertha dominica* (F.). Their reproductive potential can be reduced by lowering the temperature or drying the seed. However it would be necessary to maintain temperatures of below 20° to efficiently restrict the development of both species and temperatures of below 17°C are needed to totally stop the development of *S. oryzae*. *R. dominica* can tolerate moisture contents of as low as 10%. Howe (1965) has reviewed the literature concerning the biology of a number of storage pests and has summarised the optimal and minimal conditions for their development.

Table 2. Selected major stored seed pests: limiting conditions per population increase.

	Order and family*	Temperature (°C)		Relative humidity (%)		
		Minimum	Optimum range	Minimum	Approximately equivalent seed moisture content (%)	
Important pests of cereal seeds						
Prostephanus truncatus (Horn)**	C, B	20	27–35	40	9	Shires (1979), Bell and Watters (1982)
Rhyzopertha dominica (Fabricius)	C, B	23	32–35	30	8	
Sitophilus oryzae (Linnaeus)	C, Cu	17	27–31	60	11	
Sitophilus zeamais (Motschulsky)	C, Cu	17	27–31	60	11	
Sitotroga cerealella (Olivier)	L, Ge	16	26–30	30	8	
Trogoderma granarium (Everts)	C, D	24	33–37	1	–	
Important pests of legume seeds						
Acanthoscelides obtectus (Say)	C, Br	17	27–31	30	9	
Callosobruchus chinensis (Linnaeus)	C, Br	19	28–32	30	8	
Callosobruchus maculatus (Fabricius)	C, Br	22	30–35	10	3	
Callosobruchus analis (Fabricius)	C, Br	22	30–33	1	–	
Callosobruchus rhodesianus (Pic)	C, Br	16	25–30	?	–	Howe and Currie (1964) Giga (1982)
General feeders capable of attacking a range of susceptible and damaged seeds						
Corcyra cephalonica (Stainton)	L, Ga	18	28–32	30	8	
Cryptolestes ferrugineus (Stephens)	C, Cj	23	22–35	10	5	
Ephestia cautella (Walker)	L, P	17	28–32	25	8	
Oryzaephilus mercator (Fauvel)	C, S	20	21–34	10	5	
Oryzaephilus surinamensis (Linnaeus)	C, S	21	31–34	10	5	
Plodia interpunctella (Hubner)	L, P	18	28–32	25	8	Bell (1975)
Tribolium castaneum (Herbst)	C, T	22	32–35	1	–	
Tribolium confusum J du Val	C, T	21	30–33	1	–	

* Bo – Bostrichidae, Br – Bruchidae, C – Coleoptera, Cu – Curculionidae, Cj – Cucujidae, D – Dermestidae, Ga – Galleriinae, L – Lepidoptera, P – Pyralidae, S – Silvanidae, T – Tenebrionidae.

** *P. truncatus* is a Central American species newly introduced into E. Africa where it is posing a serious threat to maize. It may attack seed.

Sources: Howe (1965) and cited authors. The equivalent seed moisture contents for cereal pests and general feeders were based on data for maize (Gough, 1980). For *A. obtectus* they were based on data for beans (*Phaseolus vulgaris*) (Weston and Morris, 1954) and for *Callosbruchus* spp, they were based on data for cowpea (*Vigna unguiculata*) (Gough, 1978).

Table 2 is based on his summarised data (supplemented by data from other authors when necessary), and gives the minimum and optimum temperatures and the minimum relative humidity for the development of populations of a number of important seed pests. Reference to table 2 shows that, although sub-optimal temperatures could be achieved in open stores, it would be difficult in warm climates to achieve temperatures sufficiently low to totally prevent the development of pest populations. Similarly, many species are tolerant of very low relative humidities which could only be obtained in extremely dry seed. Therefore the conditions that can be obtained in open stores cannot effectively restrict insect pest development, and it is necessary to use chemicals to control them if seed is to be stored for reasonable periods of time.
The most important procedures to follow are initial disinfestation of seed, the prevention of cross-contamination from sources of infestation and long term protection.

The most successful method available for the disinfestation of seed is fumigation. This involves exposing the seed to a toxic gas that kills all pests present, including eggs and larvae boring inside seed tissues. Fumigations are carried out in gas-tight structures such as silos and bins, or under gas-proof sheeting. Small lots of seed can be fumigated in sealed oil drums or polythene sacks. All fumigations, especially large scale ones, are potentially hazardous, and should be carried out only by trained personnel who have access to necessary safety equipment. The most commonly available fumigants are methyl bromide (which is supplied as a liquified gas in cylinders and is applied to the seed through a series of tubes and nozzles) and phosphine (which is obtained from the breakdown of specially formulated tablets or packets of aluminium phosphide). Both fumigants can safely be used to disinfect seed if recommended dosages are not exceeded. Viability may be reduced if gas concentrations or exposure times are increased, especially at high moisture contents. The effects of fumigation on seed viability have been reviewed by Howe (1973), Lindgren and Vincent (1962), Page and Lubatti (1963) and Taylor (1975). For methyl bromide Monro (1970) recommends an initial dosage of 16g m^{-3} for 24 hours at temperatures over 20 °C, or 24g m^{-3} for 24 hours at 10 °–20 °C. He recommends that methyl bromide should not be used at temperatures in excess of 25 °C. The European and Mediterranean Plant Protection Organisation (1976) recommends an initial dosage of 15g m^{-3} for 24 hours for seeds with a moisture content of 14% or less, except for certain susceptible seeds such as onion (*Allium cepa* L.) and *Agrostis* spp. whose moisture content should not exceed 12%. It is stated that germination of certain seeds may be reduced by up to 5%. It is usually recommended that in the event of re-infestation, repeated fumigations with methyl bromide should not be carried out, as second and subsequent fumigations present an increased risk of damage to seed (Joubert and du Toit, 1969; Potgeiter and deBeer, 1972).

Phosphine has many advantages over methyl bromide as a seed fumigant. It is conveniently applied as aluminium phosphide in the form of large plates, packets, tablets or smaller pellets. The aluminium phosphide breaks down in the presence of atmospheric water to liberate phosphine gas. Application equipment is seldom required as pellets can be mixed with seed or placed around stacks of bags. The small pellets are especially suitable for the treatment of small quantities of seed. However, a longer exposure period is required for phosphine than for methyl bromide. Phosphine is usually regarded as having very little effect upon seed viability if recommended dosages are not exceeded, although a few studies (Joubert and du Toit, 1969; Potgeiter and de Beer, 1972) have shown some deleterious effects. Howe (1973) quotes eight authors who have shown phosphine to have little or no effect on germination. More recently, Hindmarsh (1977) has shown that fungicide-treated hybrid maize in Zambia could be safely stored for three years after fumigation with phosphine. The maize had an initial moisture content of 9% and was fumigated in polyethylene-lined jute sacks by the addition of a single pellet of aluminium phosphide, and the maize was subsequently stored in the same sacks. Monro (1970) recommends dosage at a rate of one tablet (or equivalent) per 400 litres ($2\frac{1}{2}$ tablets per m^3) for five days (at 12°–15°C) four days (at 16°–20°C) or three days (at 21°C or more). However the manufacturers of 'Phostoxin' recommend a maximum dosage of only $1\frac{1}{2}$ tablets per m^3, so it would be preferable, and possibly safer, to use their lower application rate.

Several other fumigants have been used to protect seed during storage, but they are usually not as convenient, safe or effective as methyl bromide or phosphine. Howe (1973) has briefly reviewed the use of hydrogen cyanide, ethylene dichloride mixed with carbon tetrachloride, carbon disulphide mixed with carbon tetrachloride and carbon tetrachloride alone. Fumigants provide no long term protection, and disinfested seed may be rapidly attacked by invading insects. It is important to prevent re-infestation from taking place by cleaning stores and removing and destroying all residues of spilled and broken seed which could harbour infestations of harmful insects. Cleaning should be carried out before seed is introduced into the store, and periodically during storage. Cracks and crevices in the fabric of the building which could provide refuge for pests should be repaired. All walls, floors and other surfaces can be sprayed using an environmentally-safe insecticide of low mammalian toxicity such as malathion, pirimiphos-methyl, chlorpyrifos-methyl, iodophenphos or fenitrothion at a rate of lg active ingredient (a.i.) per m^2 of surface. The outside of sacks can also be treated. Subsequently the risk of re-infestation can be reduced by re-spraying at intervals of eight weeks or more as necessary.

The re-infestation of seed may be eliminated or reduced by packing in insect proof bags or containers. Re-sealable tins or plastic containers can only be used for small lots of valuable seed on economic grounds, but tins, heat-sealed polyethylene-aluminium laminated pouches or heavy gauge polyethylene bags may be used for the commercial distribution of small quantities of seed. Such packages usually provide excellent protection against insect entry, but may occasionally be penetrated, especially when several packages placed together in an outer box are held firmly in place and provide the insects with stable inter-bag crevices into which they can crawl and begin to attack the package (Highland, 1978; Schmidt, 1980). Most larger seeds are stored in jute, cotton, woven polypropylene or paper bags, all of which provide poor protection against insects. The stitched ends of multi-wall paper sacks permit easy access to insects (especially small larvae), and woven bags can be penetrated at any point. In the food industry, methods have been developed for the incorporation of insecticides and insect repellents in the walls of bags (Highland, 1978). Such techniques may have applications in the seed industry.

The most effective means of providing long term protection to stored seed is by treatment with a contact insecticide. Although seed dressings are used extensively for the protection of the seed against pathogens and insect pests which attack after planting, most of the technology has developed in temperate countries and comparatively little effort is made to protect seed against storage pests. In his review of 1973, Howe stated that, although a few well-established insecticides were used to protect seed and there was information on their effects upon seed viability, there was little information concerning the new insecticides which are now used to protect stored food grains. There has been little progress since Howe's review, and γ HCH and DDT are still among the recommended chemicals. Both insecticides are sometimes used as seed dressings for protection against insects after planting, and therefore may have a dual function. Malathion is known to be effective against most seed pests and can be used as a dressing to protect seed during storage. Malathion has low mammalian toxicity and has the advantage that it can be used where there is a danger of seed being eaten by man, provided application rates take account of this possibility. Douglas (1975) recommends the application of malathion at a rate of 20 parts per million (ppm) a.i. and suggests that DDT can be mixed with a fungicide at a rate of 100 g of 50% wettable powder per lg of fungicide and that the fungicide should be applied at the manufacturer's recommende dosage. Malathion and DDT apparently have little or no effect upon germination (Lochner, 1962; Potgeiter and de Beer, 1972). Lindane (γ HCH) will control most storage pests at a rate of 2 ppm, but Srivastava (1958) and Grogan, Zuber, Brown, Whitehead and Stanway (1959) showed that lindane seriously affected germination four months and three months after treatment. However, both investigations used dosages of lindane intended to control pests after planting; the dosages were greatly in excess of those required to protect the seed during storage (approximately 469ppm and 1136ppm respectively).

There do not appear to have been any systematic studies of the effects of newer insecticides on the viability of seed. However, pirimiphos methyl, bromophos and bioresmethrin are all used to protect stored food grains and may be suitable for seed.

Rodents may cause considerable damage to stored seed, but in tropical countries the damage potential of insects is much greater. There are no special measures for controlling rodents attacking seeds: methods of ratproofing, trapping and poisoning developed for the protection of food stocks can be applied (Davis, 1970; Howard and Marsh, 1974).

The protection of seed against insect attack in tropical climates requires good management and constant vigilance. The measures outlined above should be incorporated into a routine and efficient pest control programme which should be designed to ensure that stores are well maintained and scrupulously clean, that seed is disinfested before pest populations can develop and that measures are taken to prevent the re-infestation of disinfested seed. Pest population build-up can be extremely rapid, and the development of new cultivars of certain crops that are highly susceptible to storage pests (Dobie, 1974; Dobie and Kilminster, 1978) can cause great problems when storing seed. Progress towards selecting more resistant cultivars (Gatehouse, Gatehouse, Dobie, Kilminster and Boulter, 1979) may eventually improve matters, but storage pests will remain a serious threat to stored seed in tropical climates.

Appendix 1. Typical climatic and building data required for the prediction of effective seed temperature.

1. Climate: Equatorial humid climate, West Africa.

t_{ao} 24 hour average = 30 °C
t_{ao} maximum = 34 °C
t_{ao} minimum = 26 °C

Radiation (Wm^{-2})

Orientation	Daily mean (I_t)	Increment (I_t^1)
South	51	104
North	157	238
East	149	6
West	149	6
horizontal (roof)	301	694
long wave (I_L)	100	

Based on solar intensities for May 22 and July 23 at latitude 5 °N assuming sky clarity factor = 1.00 applied both to direct intensities on vertical surfaces and to direct components of total horizontal intensities.

Ventilation

A minimum of five air changes per hour (constant rate) is assumed corresponding to a wind speed of 3–5m/s. This is based on calculations for natural ventilation in the IHVE Guides A2–A9 (1980).

2. Building data

(a) Dimensions: Length 40 m; Breadth 20 m; Height 6 m; Roof pitch 20 °; Orientation EW

(b) Materials: The types of materials used for the building are combinations of the following roofs and walls:

Roofs	*Description*
R1	Corrugated galvanised steel sheet 24 gauge
R2	Big six asbestos cement 3.5 mm
R3	Corrugated aluminium 0.9 mm
R4	Corrugated aluminium 0.9 mm plus 13 mm fibre board under rafters.

Walls	*Description*
W1	Big six asbestos cement 3.5 mm
W2	Light colour brick 114 mm bare
W3	Light colour brick 228 mm bare
W4	Cavity block 250 mm, rendered and painted white.

For the purpose of predicting seed temperatures the following combinations of wall and roof materials have been chosen.

R1 W1
R2 W2
R3 W3
R4 W4

and the various properties of these are shown in the table.

Table of thermal properties

Material	*Thermal properties* U	a	e	Y	f	Rso
R1	7.0	0.7	0.7	7.0	1.0	0.4
W1	5.3	0.6	0.6	5.3	1.0	0.6
R2	5.3	0.6	0.6	5.3	1.0	0.4
W2	3.6	0.5	0.5	4.0	0.8	0.6
R3	2.8	0.3	0.3	2.8	1.0	0.4
W3	2.7	0.5	0.5	4.0	0.4	0.6
R4	1.0	0.3	0.3	1.0	0.9	0.4
W4	1.2	0.3	0.9	4.0	0.3	0.6

U = transmittance ($Wm^{-2}\,°C$)
a = absorption coefficient
e = emittance
Y = admittance ($Wm^{-2}\,°C$)
f = decrement factor
R_{so} = surface resistance ($M^2\,°C\,W^{-1}$)

Sources Koenigsburger, Ingersol, Mayhew and Szokolay (1974); Petherbridge (1974) and IHVE (1980).

Note:
Moving from combinations R1 W1 to R4 W4 it can be seen that the transmittances, absorption coefficients and decrement factors all decrease in value as the quality of the materials increases. Only the admittances do not increase as would be wished to decrease temperature swings. This may be explained because no admittance values were available for the single sheet materials and the standard practice has been followed by substituting transmittance values. Substitution of the correct values could only enhance the effect of using high quality materials and would further support the arguments presented in the text.

Appendix 2. Cost data used to predict drying and storage costs

1. Drying costs
   These are based on the manufacturers performance figures for a nominal 1 tonne/hour tray drier. Variable costs were estimated at £3/hour. Fixed costs were based on an installed price of £9000 for each drier and allied equipment, annualised over a 10 year life at 10% interest.

2. Building costs
   Extra capital costs for insulation include installation costs. These are proportional to the surface areas involved and are based on information from West Africa (Joy G E (1981), private communication) and standard costing techniques (Davis, Belfield and Everest, 1981). These total costs are annualised over 20 years at 10% interest charge.
   The costs per square metre employed were

Roof tvpe	Cost (£)	Wall type	Cost (£)
R1	0	W1	9
R2	12	W2	15
R3	26	W3	25
R4	34	W4	37

It is emphasised that these are estimates which indicate the order of costs, rather than actual site costs.

References

Bell, C. H. (1975). Effects of temperature and humidity on development of four pyralid moth pests of stored products. *Journal of Stored Products Research,* **11**, 167–176.

Bell, R. J. and Watters, F. L. (1982). Environmental factors influencing the development and rate of increase of Prostephanus truncatus (Horn) (Coleoptera: Bostrichidae) on stored maize. *Journal of Stored Products Research,* **18**, 141–145.

Davis, R. A. (1970). Control of rats and mice. *UK Ministry of Agriculture, Fisheries and Food Bulletin* 181.

Davis, Belfield and Everest, Chartered Surveyors (1981). *Spon's Architects' and Builders' Price Book.* E. and F. N. Spon Ltd., 11 New Fetter Lane, London EC4P 4EE.

Delouche, J. C., Matthes, R. K., Dougherty, G. M. and Boyd, A. H. (1973). Storage of seed in sub-tropical and tropical regions. *Seed Science and Technology,* **1**, 671–700.

Dobie, P. (1974). The laboratory assessment of the inherent susceptibility of maize varieties to post-harvest infestation by *Sitophilus zeamais* Motsch. (Coleoptera, Curculionidae). *Journal of Stored Products Research,* **10**, 183–197.

Dobie, P. and Kilminster, A. M. (1978). The susceptibility of triticale to post-harvest infestation by *Sitophilus zeamais* Motschulsky, *Sitophilus oryzae* (L.) and *Sitophilus granarius* (L.). *Journal of Stored Products Research,* **14**, 87–93.

Douglas, J. E. (1975). Seed storage and packaging. In *Cereal Seed Technology,* (ed. W. P. Feistritzer), Agricultural Development Paper No. 98, FAO, Rome.

Ellis, R. H. and Roberts, E. H. (1980a). Improved equations for the prediction of seed longevity. *Annals of Botany,* **45**, 13–30.

Ellis, R. H. and Roberts, E. H. (1980b). The influence of temperature and moisture on seed viability period in barley (Hordeum distichum L.) *Annals of Botany,* **45**, 31–47.

European and Mediterranean Plant Protection Organization (1976). Methyl bromide fumigation of seeds to control stored products insects in general. *EPPO Bulletin,* **6**, 28.

Gatehouse, A. M. R., Gatehouse, J. A., Dobie, P., Kilminster, A. M. and Boulter, D. (1979). Biochemical basis of insect resistance in *Vigna unguiculata. Journal of the Science of Food and Agriculture,* **30**, 948–958.

Giga, D. P. (1982). *A comparative biology of four* Callosobruchus *species with particular reference to competition in* C. rhodesianus *and* C. maculatus. PhD. thesis, University of Reading, UK.

Gough, M. C. (1978). Moisture humidity equilibria of tropical stored produce. Part III-legumes, spices and beverages. *Tropical Stored Products Information,* **35**, 15–30.

Gough, M. C. (1980). Moisture content/relative humidity equilibria of some tropical cereal grains. *Tropical Stored Products Information* **39**, 13–18.

Grogan, C. O., Zuber, M. S., Brown, H. E., Whitehead, M. D. and Stanway, V. M. (1959). Effect of fungicides and insecticides on the germination of corn after storage. *Plant Disease Reporter,* **43**, 1132–1137.

Hindmarsh, P. S. (1977). The long term storage of hybrid maize seed in Zambia using polythene lined sacks. *Tropical Science,* **19**, 141–143.

Highland, H. A. (1978). Insect resistance of food packages – a review. *Journal of Food Processing and Preservation,* **2**, 123–130.

Howard, W. E. and Marsh, R. E. (1974). Rodent control manual. *Pest Control,* **42**, Supplement A–U.

Howe, R. W. (1965). A summary of estimates of optimal and minimal conditions for population increase of some stored products insects. *Journal of Stored Products Research,* **1**, 177–184.

Howe, R. W. (1973). Loss of viability of seed attributable to infestations of insects and mites. *Seed Science and Technology*, **1**, 563–586.
Howe, R. W. and Currie, J. E. (1964). Some laboratory observations on the rates of development, mortality and oviposition of several species of Bruchidae feeding in stored pulses. *Bulletin of Entomological Research*, **55**, 437–477.
Institution of Heating and Ventilating Engineers (IHVE) (1980). *Guides A2–A9*. Chartered Institution of Building Services, London.
Joubert, P. C. and Du Toit, D. M. (1969). The development and yield of plants derived from fumigated seeds. 1. Maize. *Phytophylactica*, **1**, 195–200.
Koeningsburger, O. H., Ingersol, T. G., Mayhew, A. and Szokolay, S. V. (1974). *Manual of tropical housing and building*, Vol. I. Longmans, London.
Lindgren, D. L. and Vincent, L. E. (1962). Fumigation of food commodities for insect control. *Advances in Pest Control Research*, **5**, 85–152.
Lochner, E. H. W. (1962). Giftigheidgehalte van Kontakgifstawwe vir graaninsekte wanneer gemeng met graan. *Technical Communication of the South African Department of Agriculture Technical Service*, No. 9.
Monro, H. A. U. (1970). *Manual of fumigation against insects*. Agricultural Development Paper N. 79, FAO, Rome.
Page, A. B. P. and Lubatti, O. F. (1963). Fumigation of insects. *Annual Review of Entomology*, **8**, 239–264.
Petherbridge, P. (1974). Limiting the temperatures in naturally ventilated buildings in warm climates. *Building Research Establishment. Current Paper*, Feb. 1974.
Potgeiter, H. V. and De Beer, P. R. (1972). The effect of certain contact insecticides and fumigants on the germination and yield after growth of stored grain seed. *Proceedings of the International Seed Testing Association*, **37**, 941–951.
Roberts, E. H. (1961). Viability of cereal seed for brief and extended periods. *Annals of Botany*, **25**, 373–380.
Schmidt, H. U. (1980). Über die schutzwirkung von packstoffen und fertigpackungen gegen befall durch voratsschädlinge. *Getreide Mehl und Brot.*, **34**, 215–27.
Shires, S. W. (1979). Influence of temperature and humidity on survival, development period and adult sex ratio in *Prostephanus truncatus* (Horn) (Coleoptera: Bostrichidae). *Journal of Stored Products Research*, **16**, 147–150.
Srivastava, B. K. (1958) Influence of certain insecticidal seed treatments on wheat. *Madras Agricultural Journal*, **45**, 10–12.
Taylor, R. W. D. (1975). The storage of seeds. *Tropical Stored Products Information*, **30**, 23–34
Weston, W. J. and Morris, H. J. (1954). Hygroscopic equilibria of dry beans. *Food Technology-Champaign*, **8**, 353–355.

Ellis, R. H., Hong, T. D. and Roberts, E. H. (1983), *Seed Sci. & Technol.*, **11**, 77–112

Procedures for the safe removal of dormancy from rice seed

R. H. ELLIS, T. D. HONG and E. H. ROBERTS

Department of Agriculture and Horticulture, University of Reading, Earley Gate. Reading RG6 2AT, UK

(Accepted August 1982)

Summary

Techniques for dormancy removal and recommended procedures for germination tests in rice are reviewed. It is argued that most existing recommendations are not entirely satisfactory. Consequently investigations were undertaken on a wide range of genetic material: 14 cultivars representing *Oryza glaberrima* and the three ecogeographic races of *O. sativa* (*indica, japonica* and *javanica*). The factors examined included the effects of temperature and moisture content on dry after-ripening, pre-soaking treatments at various temperatures, various chemicals (GA_3, $GA_{4/7}$, KCN, NaN_3, thiourea, 2-mercaptoethanol, HNO_3 and H_2O_2) and a wide range of alternating and constant temperature regimes during the germination test. No single satisfactory treatment was found which was applicable to all taxonomic groups. The procedure recommended by ISTA for removing dormancy (pre-soak for 24–48 hours in water at 40 °C) is questioned as not being very effective and potentially damaging to some cultivars. In *japonica* cultivars (which usually show only slight dormancy) a diurnal temperature alternation of 30°/20 °C is satisfactory but, contrary to existing ISTA prescriptions, it is preferable for the higher temperature to be applied for the longer period (16 hours). For cultivars of *indica, javanica* and *O. glaberrima,* however, an alternation of 34°/11 °C is preferable, again with 16 hours at the higher temperature. In the case of dormant *O. glaberrima* cultivars this treatment alone is insufficient, but a pre-treatment in 0.1M HNO_3 for 24 hours followed by a further 24 hours in 0.25M H_2O_2 (both at 20 °C) and germination at 34°/11 °C in 0.01M 2-mercaptoethanol appears to be very satisfactory. The moisture content of dry stored seed of *O. glaberrima* affected the rate of after-ripening. At high temperatures loss of viability occurred before complete loss in dormancy. After-ripening at lower temperatures resulted in complete loss in dormancy, but this is not a satisfactory procedure for breaking dormancy in *O. glaberrima* because the period of treatment required is too long.

Résumé

Procédés pour la levée sans risque de la dormance des semences de riz

On passe en revue les techniques de levée de dormance et les procédés recommandés pour les essais de germination chez le riz. Il est démontré que la plupart des recommandations existantes ne sont pas entièrement satisfaisantes. Par conséquent, des recherches ont été entreprises sur une gamme étendue de matériel génétique: 14 cultivars représentant *Oryza glaberrima* et les trois races écogéographiques d' *O. sativa* (*indica, japonica* et *javanica*). Les facteurs examinés comprennent les effets de la température et de la teneur en eau sur la post-maturation des grains secs, les traitements de pré-trempage à diverses températures, divers produits chimiques (GA_3, $GA_{4/7}$, KCN, NaN_3, thiourée, 2-mercaptoéthanol, HNO_3 et H_2O_2) et une gamme étendue de régimes de températures alternées et constantes pendant l'essai de germination. On n'a trouvé aucun traitement satisfaisant unique qui soit applicable à tous les groupes taxonomiques. La méthode recommandée par l'ISTA pour lever la dormance (pré-trempage pendant 24–48 heures dans l'eau à 40 °C) est contestée comme n'étant pas très efficace et pouvant causer des dégâts chez quelques cultivars. Dans les cultivars de *japonica* (qui ne montrent habituellement qu'une légère dormance) une alternance de température diurne de 30°/20 °C est satisfaisante, mais contrairement aux prescriptions existantes de l'ISTA, il est préférable que la température la plus élevée soit appliquée pendant la période la plus longue (16 heures). Pour les cultivars d'*indica, javanica* et *O. glaberrima*, cependant, une alternance de 34°/11 °C est préférable, avec aussi 16 heures à la température la plus haute. Dans le cas de cultivars dormants de *O. glaberrima*, ce traitement seul est insuffisant, mais un pré-traitement dans HNO_3 à 0.1M pendant 24 heures suivi de 24 heures dans H_2O_2 à 0.25M (les deux à 20 °C) et une germination à 34°/11 °C dans du 2-mercaptoéthanol à 0.01M se révèlent très satisfaisants. La teneur en eau des semences de *O. glaberrima* conservées au sec affectait la vitesse de post-maturation. A des températures élevées, la perte de viabilité survenait avant la levée totale de la dormance. La post-maturation à des températures plus basses a entraîné une levée complète de la dormance, mais ce n'est pas un procédé satisfaisant pour lever la dormance chez *O. glaberrima* parce que la période de traitement requise est trop longue.

Zusammenfassung

Verfahren zur sicheren Beseitigung der Keimruhe bei Reissamen

Es wird ein Überblick über Techniken zur Beseitigung der Keimruhe und über empfohlene Methoden für Keimprüfungen bei Reis gegeben. Es wird gefolgert, daß die meisten der bestehenden Empfehlungen nicht ganz zufriedenstellend sind. Infolgedessen wurden Untersuchungen in einem ausgedehnten Bereich genetischen Materials vorgnommen, nämlich an 14 sorten von *Oryza glaberrima* und drei ökogeographischen Rassen von *O. sativa* (*indica, japonica* und *javanica*). Zu den geprüften Faktoren gehörten die Einflüsse von Temperatur und Feuchtigkeitsgehalt bei trockener Nachreife, Vorquellungsbehandlungen bei verschiedenen Temperaturen, verschiedene Chemikalien (GA_3, $GA_{4/7}$, KCN, NaN_3, Harnstoff, 2-Merkaptoäthanol, HNO_3 und H_2O_2) und ein weiter Bereich von Wechsel- und konstanten Temperaturen während der Keimprüfung. Man fand keine einzelne Behandlung, welche bei allen taxonomischen Gruppen anwendbar war. Das von der ISTA empfohlene Verfahren zur Beseitigung der Keimruhe (24–28 Stunden Vorquellung in Wasser von 40 °C) wird als nicht sehr wirksam und möglicherweise schädigend bei manchen Sorten in Frage gestellt. Bei *japonica*-Sorten (welche gewöhnlich nur eine leichte Keimruhe aufweisen) ist eine tägliche Wechseltemperatur von 30°/20 °C ausreichend, aber im Gegensatz zu den derzeitigen ISTA-Vorschriften ist es besser, die höhere Temperatur für eine längere Zeitspanne (16 Stunden) anzuwenden. Bei den Sorten von *indica, javanica* und *O. glaberrima* jedoch ist ein Wechsel von 34°/11 °C besser, wieder mit 16 Stunden bei der höheren Temperatur. Im Falle von keimruhenden *O. glaberrima*-Sorten ist diese Behandlung allein nicht ausreichend, aber eine Vorbehandlung in 0,1 M HNO_3 für 24 Stunden, gefolgt von weiteren 24 Stunden in 0,25 M H_2O_2 (beide bei 20 °C) und eine Keimungsdurchführung bei 34°/11 °C in 0,01 M 2-Merkaptoäthanol scheint sehr zufriedenstellend zu sein. Der Feuchtigkeitsgehalt des trocken gelagerten Saatguts von *O. glaberrima* beeinflußte den Grad der Nachreife. Bei höheren Temperaturen trat ein Verlust an Lebensfähigkeit vor dem völligen Verlust der Keimruhe auf. Nachreife bei niedrigeren Temperaturen führte zum völligen Verlust der Keimruhe. Bei *O. glaberrima* ist das aber kein zufriedenstellendes Verfahren zur Brechung der Keimruhe, weil die erforderliche Behandlungszeit zu lang ist.

Introduction

Seed dormancy in rice (*Oryza* spp.) is both problematic and advantageous. It is problematic for post-harvest seed testing (Delouche and Nguyen, 1964), in the production of two or three crops a year (Ghosh, 1960) and in plant breeding (Roberts, 1961a). It is advantageous, however, in avoiding viviparous germination, for example in tropical cultivars grown during the monsoon season (Agrawal and Nanda, 1969).

There is a large literature on rice seed dormancy. Roberts and Smith (1977) provided a list of 22 treatments which bring about loss of dormancy within seed of 135 graminaceous species, but their list for rice was somewhat incomplete. Consequently we have taken this opportunity to provide a fuller list below of treatments which have been reported to reduce seed dormancy in the Gramineae, together with references to their reported activity in rice (stimulatory, unless noted otherwise).

Removal or puncture of seed covering structures (lemma-palea and/or pericarp and testa)
Often the effects are large though they may be only slight in some seed lots, particularly those of *O. glaberrima* Steud. (Oka and Tsai, 1955; Roberts, 1961b; Nakamura, 1963; Delouche and Nguyen, 1964; Tseng, 1964; Sikder, 1967; Misra and Misro, 1970; IRRI, 1972, 1973; Ikehashi, 1973; Sugawara, 1973).

High oxygen tension
Increased oxygen partial pressures applied either during storage (Roberts, 1962; Yasue and Asai, 1970) or in the germination test (Sikder, 1967; Hayashi and Morifuji, 1972; Hayashi, 1980) is stimulatory, even though non-dormant rice seed can germinate (slowly) in little or no oxygen.

High nitrogen tension
Paradoxically high nitrogen pressures in seed storage have also been reported to reduce dormancy (Yasue and Asai, 1970); but Roberts (1962) reported that storing seeds at atmospheric pressure in nitrogen delayed loss of dormancy and that the germination percentage of partially dormant seed is lower in nitrogen than in air.

Oxidising agents
Hydrogen peroxide is often quite effective (Roberts, 1964a; Yasue, 1973) and sometimes other oxidising agents, e.g. sodium hypochlorite (Mikkelsen and Sinah, 1961; Delouche and Nguyen, 1964); although Roberts (1964a) found that calcium hypochlorite and a number of other oxidising agents had no effect.

Nitrate or nitrite
These ions are usually stimulatory (Roberts, 1963a; IRRI, 1968; Agrawal and Nanda, 1969; Subramoney and Abraham, 1969; Jalote and Vaish, 1976). The property seems to be associated with the electron accepting role of these ions rather than their nitrogen content since the other reduced nitrogenous compounds, e.g. ammonium ions, urea and amino acids, have little or usually no effect (Roberts, 1963a) unless they have other special properties, e.g. some of the respiratory inhibitors mentioned below.

Alternative electron acceptors
Methylene blue has a slight stimulatory effect (Roberts, 1964a) and this tends to support the interpretation of the effect of nitrate ions.

Respiratory inhibitors
The activity of most of the agents above suggests that some oxidative activity is in-

volved in loss of dormancy. However this activity does not appear to be associated with conventional respiration since inhibitors of terminal oxidase, e.g. potassium cyanide, sodium azide, hydroxylamine, hydrogen sulphide and carbon monoxide are stimulatory to varying extents (Roberts, 1964a, 1964b). However, unlike some other species, inhibitors of the Krebs cycle or of glycolysis had no effect (Roberts, 1964a) (although they could have other effects, since the end-products of respiration mentioned below also act as respiratory inhibitors).

End-products of fermentation

Ikeda (1963) reported that ethyl alcohol was stimulatory. However lactate and acetate had no effect (Roberts, 1963b). Carbon dioxide applied to wet seeds can also be stimulatory (Tseng, 1964) but in dry storage carbon dioxide (as compared with air storage) delays loss of dormancy (Roberts, 1962).

Sulphydryl compounds

Some sulphydryl compounds are stimulatory (Nakamura, 1963; Roberts, 1963b, 1964a; Tseng, 1964; Murty and Raghavaiah, 1966) but others are not (Roberts, 1963b, 1964a).

Hydrogen ions

Pretreatment in acids, e.g. sulphuric acid (Murty and Raghavaiah, 1966; Sikder, 1967; Sugawara, 1973) or buffer solutions at pH values of pH 2.0 or below stimulates the loss of dormancy; at pH 1.1 the effect can be quite marked (Roberts, 1963a). Nitric acid can be more effective than other acids because the nitrate ion also has an effect (Roberts, 1963a).

Growth hormones

Some effect is usually shown by gibberellic acid (Nakamura, 1963; Roberts, 1963b; IRRI, 1972), a lesser effect by kinetin and sometimes a marginal effect is shown by indoleacetic acid (Roberts, 1963b). There was no evidence of interactions between these hormones when combined (Roberts, 1963b). Ethylene has also been shown to have some effect (Delouche and Nguyen, 1964).

Temperature and water

The term after-ripening usually refers to the process whereby air-dry seeds gradually lose dormancy with the progress of time, and the higher the temperature the more rapid the loss of dormancy (Oka and Tsai, 1955; Dore, 1955; Roberts, 1961a, 1962, 1965; Jenning and de Jesus, 1964; Delouche and Nguyen, 1964; Murty and Raghavaiah, 1966; IRRI, 1968; Agrawal and Nanda, 1969; Ikehashi, 1970; Hayashi and Morifuji, 1972; Yasue, 1973; Jalote and Vaish, 1976; Chang, 1976a).

Soaking in water at room temperature (Delouche and Nguyen, 1964; Yasue, 1973), or at 40 °C (Delouche and Nguyen, 1964; Jalote and Vaish, 1976), or at 40 °C with applied oxygen (Hayashi, 1977) can be effective in breaking dormancy; but in other cases the effect of soaking at 27 °C is small and sometimes non-existent (Roberts, 1962).

Surprisingly, for a tropical species, soaking at 3 °C (stratification) also stimulates loss of dormancy (Roberts, 1962).

Alternating temperatures applied to imbibed seeds during germination tests can promote loss in dormancy in rice (Nakamura, 1963; Chaudhary and Ghildyal, 1969; Hayashi and Morifuji, 1972; Singh, Singh and Khanna, 1973).

Light

Light applied during storage (Yasue and Asai, 1970) or during the germination test (Roberts, 1962) has been reported to improve the germination of dormant seeds slightly; but Delouche and Nguyen (1964) found no effect whilst Nakamura (1963) reported an inhibition of germination.

Miscellaneous compounds

Quinones, ascorbic acid and chelating compounds have sometimes been reported to have effects in graminaceous species but no positive reports on rice have been found and Roberts (1964b) found no effects of quinones or chelating agents. Murty and Raghavaiah (1966) reported that mercuric chloride was stimulatory but Roberts (1963a) found that it delayed breaking of dormancy in serial germination tests on stored seeds.

Although there are some contradictions, there is considerable agreement on the main factors which affect dormancy in rice. But, despite the long list of promotory agents, it is unusual for any single dormancy-breaking agent to promote the germination of all dormant seeds within a lot. For example, in much of the work reported by Roberts full germination only occurred when the various promotory agents were applied to seed that had already been partly, but incompletely, after-ripened (Roberts, 1963a, 1963b, 1964a). Consequently, although some workers have attempted to produce practical procedures for breaking rice seed dormancy that can be used to allow viability to be estimated by germination tests or to promote full plant establishment immediately post-harvest, there is little evidence to suggest that they are satisfactory standard procedures which can be applied in all cases. In seed testing laboratories a 14 day test either in a diurnal alternating temperature regime of 20°/30 °C (the higher temperature being applied for eight hours per day) or at a constant 30 °C is prescribed (ISTA, 1976b, 1981). Prior to this test ISTA recommend one to two days soaking in water at 40 °C to break dormancy (ISTA, 1976b). Where the water soak treatment at 40 °C is insufficient, soaking in 0.1% ethylene chlorohydrin or 0.25% sodium hypochlorite for 24 hours has been suggested (Delouche and Nguyen, 1964). A further alternative pretreatment is to soak seed in 0.1M HNO_3 for 23 hours (Agrawal and Nanda, 1969) or 16 to 24 hours (IRRI, 1968), and it appears that this procedure has a further advantage as a treatment against bacterial blight (Subramoney and Abraham, 1969). To enable seedlings to be raised from dormant rice seed (for example in plant breeding programmes) it has been suggested that the seeds be after-ripened by storing at 47 °C for seven days (Carpenter and Roberts, 1962). To reduce the risk of loss of viability occurring as a result of this treatment, it has been recommended that the treatment only be applied to seed already dried to 11% moisture content and to after-ripen in open containers to allow the seed to dry further (Carpenter and Roberts, 1962). This was suggested because reducing seed moisture content reduces the rate of loss in viability in rice seed (Roberts, 1961c) but, apparently, not the rate of after-ripening (Roberts, 1962). Similar procedures have been suggested by others (Chang, 1976a; Jenning and de Jesus, 1964).

In order to develop methods of breaking dormancy which are widely applicable it is necessary to ensure that experiments are undertaken on a wide genetic range of material. Wide applicability is not only relevant to routine seed testing but is particularly important to plant breeders who often wish to sow successive generations in rapid sequence and are therefore confronted with dormancy in its most pronounced form, i.e. before there has been any significant after-ripening. Under these circumstances unless dormancy is removed by physiological techniques there will be a tendency in segregating material for genetic selection for little seed dormancy which, in turn, will lead to the development of cultivars with a tendency to sprout in the head. It is also of special importance to seed banks for genetic conservation where, in order to obtain maximum longevity, it is necessary to store good quality seeds (Ellis and Roberts, 1980); and this means placing the accessions in cold store as soon as possible after harvest. Thus when using germination tests to estimate initial viability, and also in monitoring viability in store subsequently, dormancy is likely to be a problem since dormancy tends to be maintained for very long periods when dry seeds are stored at sub-zero temperatures.

Accordingly in this study we have examined a wide range of genetic material. There are two species of cultivated rice: *Oryza sativa* L. is the main rice of commerce which originated in S.E. Asia; *O. glaberrima* Steud. originated in West Africa and is of local importance there. *O. sativa* is divided into three ecogeographic races (sometimes considered to be sub-species): *indica* which is tropical; *japonica* which is sub-tropical; and *javanica* which is of more limited distribution, particularly in the area centred on Indonesia, and which in some respects has characters intermediate between *indica* and *japonica* (Chang, 1976b). Of these, presumably because they have been subjected to less selection pressure, *O. glaberrima* cultivars are generally the most dormant, followed by *indica, javanica* and *japonica* cultivars, the latter being the least dormant. However it would appear that dormancy in all is of a similar type in that they all respond in a similar manner, though to different degrees, to dormancy-breaking treatments. Although, apart from their possible use in breeding programmes, *O. glaberrima* cultivars are of only limited interest in W. Africa, we included four cultivars in this study since we felt that, because of their pronounced dormancy, they provide very stringent test material for dormancy-breaking techniques. Ten cultivars of *O. sativa* have also been used, four *indica,* four *japonica* and two *javanica* (table 1).

We report, in chronological order as far as possible, a series of investigations in which we originally tested the existing recommended dormancy-breaking treatments for their efficacy and subsequently attempted to develop a more suitable standard procedure which could be safely applied to promote the germination of all dormant rice seeds, without reducing viability.

Table 1. Information on rice seed lots used in these investigations.

Cultivar	Breeders number/name	Species (sub-species)	Supplier	Original source	1,000[1] seed weight, g	Initial[2] moisture content, %	Germination,[3] %	Viability,[4] %	Apparent[5] dormancy, %
A	C13H3	*Oryza sativa, indica*	RRS[6]		31	11.3	89	96	5–10
B	Chainung 242	*Oryza sativa, japonica*	RRS	Taiwan	23	11.5	93	100	5–10
C	Kolokalan	*Oryza glaberrima*	RRS	Sierra Leone	19	11.6	88	82	nil
D	CP4	*Oryza sativa, indica*	RRS		18	11.5	66	99	30–35
E	ROK3	*Oryza sativa, indica*	RRS	Sierra Leone	24	11.4	65	76	10–15
F	Fujisaka 5	*Oryza sativa, japonica*	IRRI[7]	Japan	25	11.7	98	97	nil
G	102948	*Oryza glaberrima*	IRRI	Mali	29	11.7	2	95	90–95
H	102949	*Oryza glaberrima*	IRRI	Mali	32	11.8	2	95	90–95
I	102950	*Oryza glaberrima*	IRRI	Mali	27	11.2	1	96	90–100
J	Tainan 3	*Oryza sativa, japonica*	IRRI	Taiwan	24	11.2	97	99	< 5
K	Tainan 7	*Oryza sativa, japonica*	IRRI	Taiwan	22	11.0	98	99	< 5
L	BPI76NS	*Oryza sativa, indica*	IRRI	Philippines	22	10.9	97	94	nil
M	Hawaru Batu	*Oryza sativa, javanica*	IRRI	Indonesia	29	11.3	96	97	< 5
N	Ijo Gading	*Oryza sativa, javanica*	IRRI	Indonesia	24	9.0	63	81	15–20

[1]at initial moisture content; [2]fresh-weight basis; [3]normal germination, 14 day test of 400 seeds at 20°/30°C (16 hours/8 hours); [4]Tetrazolium test on 100 moist bisected grains, soaked in 0.5% solution at 40°C for 2 hours; [5]difference between germination and viability; [6]Rice Research Station, Sierra Leone; [7]International Rice Research Institute, Philippines.

Materials and methods

Fourteen rice seed lots with varying levels of dormancy were obtained from the Rice Research Station, Sierra Leone and the International Rice Research Institute (IRRI), the Philippines (table 1). Seeds were hermetically stored at the original moisture content (table 1) and –20 °C between receipt and the start of each investigation (in order to prevent dormancy being lost as a result of the after-ripening which would occur at higher temperatures). Only a few of these lots were used for many of the experiments described here, but in the final experiment all 14 lots were tested.

Seed moisture content was determined by the method prescribed by ISTA (1976a, 1976b); viz., two replicates of 4.5–5 g of ground seed were dried for two hours in a ventilated oven maintained at 130°–133 °C and moisture content determined on the fresh-weight basis as the loss in weight during drying expressed as a percentage of the original sample weight.

Originally germination tests were carried out in moist, rolled paper towels (Bowater T20). However, it was subsequently observed that the use of these towels resulted in abnormal radicle growth of a small proportion of seeds of some seed lots that had

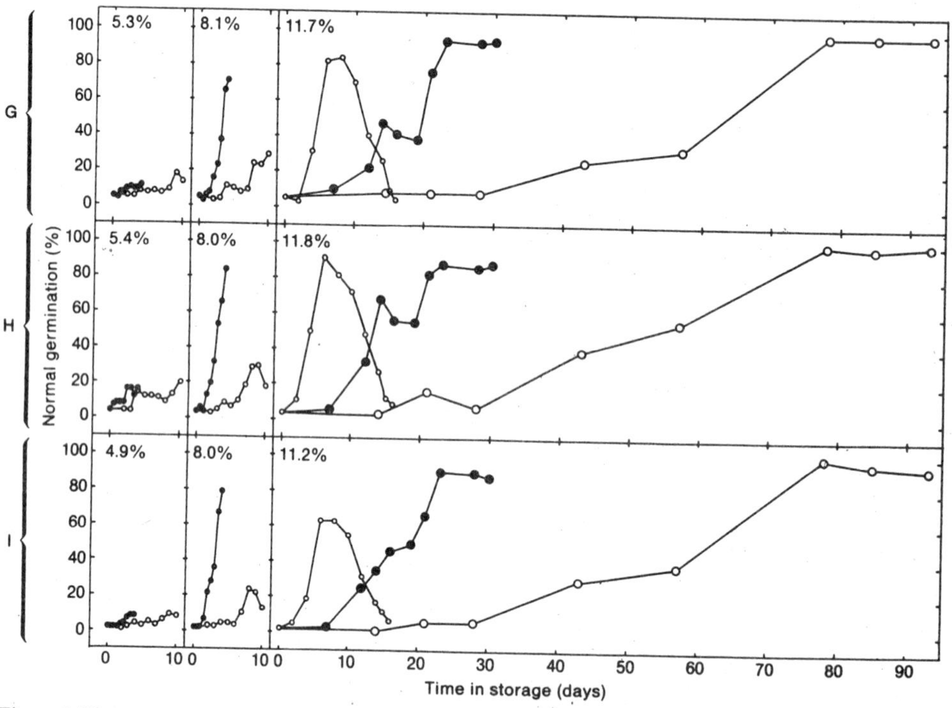

Figure 1. The percentage germination of seed of three rice cultivars (G, H, I; *O. glaberrima*) plotted against time in storage at the moisture contents shown and 30° (○), 40° (●), 50° (o) or 60 °C (•). Each point is the result of a 200 seed, 14 day germination test between moist rolled paper towels at 20°/30 °C.

been subjected to soaking pre-treatments. Consequently in later germination tests 9 cm petri dishes containing two discs of white filter paper (Whatman, grade 181) wetted with 4.5 ml of distilled water, or the appropriate solution, were used. The number of seeds which germinated normally and abnormally were recorded separately. In general results for normal germination only are provided. However, in those cases where it is suspected that the use of the towels affected normal germination, the results for total germination (that is the total numbers of seeds germinating either normally or abnormally) are provided. Except where noted otherwise, the results for both criteria of germination responded similarly to treatment. Due to the restricted height within the petri dishes, seedlings were evaluated at a comparatively early stage in seedling growth.

For clarity and to avoid repetition, the detailed methods of each investigation are provided with the results, usually with the figure captions. Some general points, however, are noted below.

In the after-ripening investigation seeds were hermetically stored in 100 mm square laminated aluminium foil packets with the following specification: an outer layer of 17 g/m^2 Melinex; 4 g/m^2 lacquer; a middle layer of 33 g/m^2 aluminium foil; 4 g/m^2 lacquer; and an inner layer of 63 g/m^2 polyethylene. Seed were dried to 8% or 5% moisture content in desiccators over (regularly regenerated) silica gel at 20 °C.

All soaking pre-treatments were done at 20 °C in distilled water, or the appropriate solution, equilibrated to 20 °C, unless noted otherwise. These are sometimes referred to as pre-applied treatments. Chemical treatments applied throughout the germination test are referred to as co-applied treatments.

The two-dimensional temperature gradient plate used to investigate the response of rice seed germination to constant and alternating temperatures has been described elsewhere (Ellis, Hong and Roberts, 1982). All investigations reported here were carried out on a daily alternating cycle of 16 hours/8 hours at various temperatures between 11°–38 °C. All alternating temperature treatments are reported as (the temperature applied for 16 hours)/(the temperature applied for eight hours). For example, 20°/30 °C means that 20 °C was maintained for 16 hours per day and 30 °C was maintained for eight hours per day.

Results and Discussion

The effect of temperature and moisture on 'air-dry' seed ('dry after-ripening')

All three storage factors investigated (treatment period, storage temperature, and seed moisture content) influenced the proportion of seeds which germinated in subsequent tests, but the pattern of loss in dormancy within each separate environment was very similar for the three cultivars (figure 1). Increasing the temperature and period of storage at constant moisture contents increased the percentage germination of the previously dormant seed of these *O. glaberrima* cultivars (figure 1) in the manner demonstrated and quantified for cultivars of *O. sativa* and *O. glaberrima* by Roberts (1962, 1963c, 1965). The temperature coefficient, Q_{10}, for the change in rate of loss in dor-

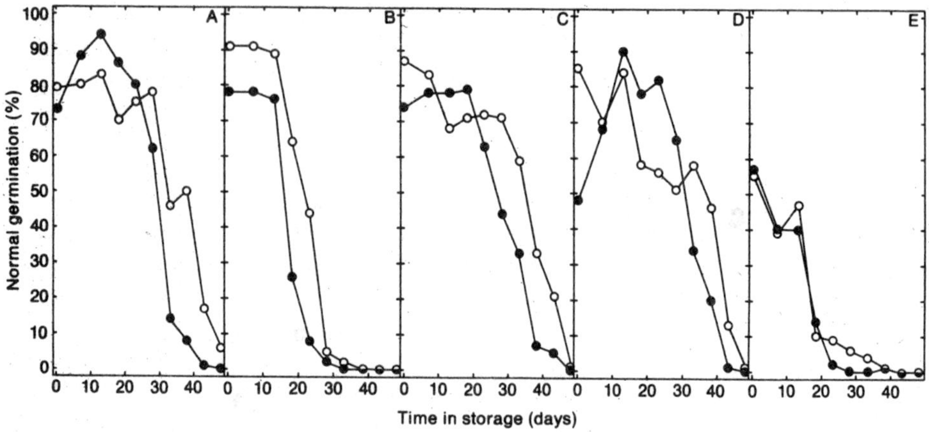

Figure 2. The percentage germination of seed of five rice cultivars in serial tests after 0–49 days hermetic storage at 40°C with 15.1% (cultivar A), 14.8% (cultivar B), 15.1% (cultivar C), 15.0% (cultivar D) or 15.1% (cultivar E) moisture content. One hundred seeds were tested in moist rolled paper towels for 14 days at 20°/30°C either without pre-treatment (open circles) or after two days soaking in water at 40°C (closed circles).

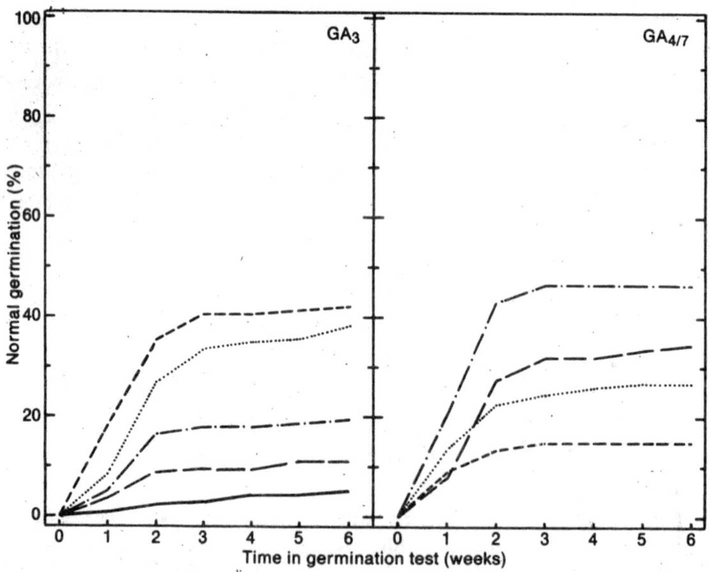

Figure 3. Germination progress curves for 100 seeds of rice cultivar G (*O. glaberrima*) tested in petri dishes on moist filter paper at 20°/30°C following soaking in water (——), or 10^{-4}M (— —), 10^{-3}M (—·—), 5×10^{-3}M (.........), or 10^{-2}M (------) GA_3 or $GA_{4/7}$ for one day (100 seeds in 12.5 ml).

mancy (reciprocal of the number of days storage required for germination to rise to 50%) was 3.5 for seed at 11% moisture content between 30°–50 °C. The value of Q_{10} did not differ significantly between the three cultivars ($P > 0.1$) and is similar to those determined previously over similar temperature ranges: 3.1 (Roberts, 1965) and 3.4 (Roberts, 1962) for rice and 3.9 (A. Minchin, unpublished) for barley (*Hordeum vulgare* L.). Using the data for the time taken for germination to rise to 20% at 8% moisture content and between 50°–60 °C provided an estimate of 3.4 for Q_{10} (all three cultivars). Thus the value of Q_{10} appears to be stable over the range 30°–60 °C, in contrast to the increase in Q_{10} values that has been observed for loss in viability of seed of many species over a wide temperature range (Ellis and Roberts, 1980, 1981).

As has already been noted, seed moisture content also affected the pattern of loss in dormancy. At 5% moisture content little loss in dormancy occurred over four days at 60 °C, whereas at the same temperature but 8% moisture content all three cultivars showed a substantial increase in germination (figure 1). A similar difference can be seen between seed at 8% or 11% moisture content over 10 days at 50 °C (figure 1). Reduction in seed moisture content from 11% to 8%, at a constant temperature, increased the storage period required for a given proportion of seeds to lose dormancy by a factor of approximately 2.5.

Clearly, then, the rate of after-ripening in rice seed is influenced substantially by seed moisture content; but this contrasts with earlier work where it was concluded that moisture content had little or no effect (Roberts, 1962). However, in that investigation only moisture contents over the range 12.0% to 14.5% were investigated and the results reported here (figure 1) for slightly lower moisture contents are unequivocal. Moreover, they agree with results over similar moisture content ranges for seed of other species, e.g. *Draba verna* L. (Baskin and Baskin, 1979), *Avena ludoviciana* Dur. (Quail and Carter, 1969), *Brassica cernua* Matsum., *B. japonica* (Thunb.) Sieb. and *B. napus* L. (Tokumasu, Kato and Yano, 1975). The conclusion that moisture content influences the rate of after-ripening in rice seed has an important practical implication: it suggests that the after-ripening method of breaking dormancy proposed by Carpenter and Roberts (1962) (see introduction) may not be an optimum procedure because the seeds are allowed to dry during the treatment.

Within the range of treatments investigated here only 78 days storage at 30 °C with 11% moisture content was almost completely effective in breaking the dormancy of seed of all three cultivars. At 50 °C, and possibly at 40 °C, although the rate of loss in dormancy was higher, loss in viability began to occur before complete loss in dormancy (figure 1). This difference in the relative rates of loss of dormancy and viability between 30° and 50 °C is to be expected since the Q_{10} for rate of loss of rice seed viability, 4.9 (Roberts, 1961c), is greater than that for loss in rice seed dormancy, 3.4 (Roberts, 1962), 3.1 (Roberts, 1965) and 3.4–3.5 (figure 1), over this temperature range. Thus, when choosing a temperature at which to after-ripen rice seed, the lower the temperature the less likely will loss in viability be confounded with loss in dormancy. However, although it is suitable for breaking dormancy in these three cultivars, a temperature of 30 °C is unlikely to prove suitable as a practical technique to either plant breeders or seed testers: first because the storage period required for complete loss in dormancy of very dormant lots is too long; secondly because the increase in germination resulting from loss in dormancy may be counteracted by a reduction in germination resulting from loss in viability for poor quality seed lots.

The three cultivars investigated here were all *O. glaberrima* and show much more pronounced dormancy than is typical of *O. sativa*. For example in a previous investigation the storage time at about 11% moisture content and 32 °C for two *O. glaberrima* cultivars to lose 50% dormancy was 80 and 90 days, whereas the comparable range for four *indica* cultivars was 15 to 30 days (Roberts, 1965). In that investigation the treatment time to achieve complete loss of dormancy at 47 °C was seven days or less for the *indica* cultivars but the *O. glaberrima* cultivars required 30 to 60 days. Under these conditions the *indica* cultivars suffered no loss of viability. Dry after-ripening at 47 °C could still therefore be considered to be a suitable technique for some purposes for less dormant cultivars of *O. sativa,* but the original technique suggested (Carpenter and Roberts, 1962) should be modified so that the seeds are not allowed to dry to below 11% moisture content. But in poor quality seed lots there could be some danger of some loss of viability and the increase in mutation which accompanies it (Roberts, 1972). Furthermore the technique is clearly unsuitable for deeply dormant cultivars of *O. glaberrima* and therefore it cannot be recommended as a suitable universal dormancy-breaking treatment for rice in general.

The effect of soaking in hot water on imbibing seed

Soaking in water at 40 °C for 1–2 days is the method of breaking rice seed dormancy recommended by ISTA (1976b) for use in seed testing stations. Others (see introduction) have also advocated the use of similar procedures. In considering this treatment we had two basic questions. First, did the use of this procedure result in damage to poor quality seed lots? Secondly, was it effective in breaking dormancy?

To answer the first question methodically we investigated whether the regular use of this technique influenced the pattern of rice seed survival curves in serial germination tests. The results for *japonica* cultivar B confirmed our original suspicions: a deleterious effect of the soaking treatment was observed following all storage periods (figure 2). However, the deleterious effect on germination was not consistent throughout the storage of the remaining four cultivars. In *indica* cultivar E the pattern of decline in germination during storage was very similar for both the control and the hot water soaking treatments until after 23 days storage when there was a small but consistent deleterious effect of soaking. There was no evidence in this cultivar of any promotion of germination for the small proportion of dormant seed (table 1) following soaking. The variable results for the normal germination of *indica* cultivar D were due to a large, variable proportion of abnormal seedlings in these tests; throughout storage results for total germination were similar following both treatments. Following 20–30 days storage the germination of the *indica* cultivar A and the *O. glaberrima* cultivar C was reduced by the soaking treatment. However, prior to this there was some evidence of a loss in dormancy resulting from the soaking treatment of seed of cultivar A: but only when this treatment was combined with a short period of dry storage.

Only a very small proportion of the dormant seeds of *O. glaberrima* cultivar G (table 1) were promoted to germinate by one or two days soaking in water at 40 °C, the treatment only raising germination from 3% (control treatment) to 9% when tested (100 seeds per treatment on moist filter paper) at 20°/30 °C for 42 days. Moreover, comparison with the results following soaking in water at 20 °C (figure 3) suggests that at least a part of this small effect was not dependent upon the higher water temperature.

Thus we conclude that the soaking of rice seed in water at 40 °C for one to two days is not an appropriate dormancy breaking treatment. First it is not sufficiently effective in promoting the germination of dormant seed. Secondly the treatment can be deleterious to the germination of non-dormant seed.

The effect of soaking in gibberellic acid solutions on imbibing seed

Gibberellic acid (GA) has been reported to promote the germination of dormant seed of at least 23 graminaceous species (Roberts and Smith, 1977), including rice (Nakamura, 1963; Roberts, 1963b), but it seldom stimulates the germination of all dormant seeds within a lot. Here pre-applied GA_3 and $GA_{4/7}$ stimulated the germination of some of the dormant seed within the very dormant *O. glaberrima* cultivar G (figure 3). With GA_3 the highest proportion of seeds (42%) germinated at the highest concentration applied (10^{-2}M), but with $GA_{4/7}$ the highest proportion of seeds (47%) germinated at an intermediate concentration (10^{-3}M) and applications of higher concentrations progressively reduced final germination. Although pre-applied GA_3 and $GA_{4/7}$, at 10^{-2}M and 10^{-3}M respectively, promoted the germination of about half of the dormant seed (see table 1), these treatments alone were unable to stimulate full germination of the dormant seed population. This conclusion is in broad agreement with the results of IRRI (1972) on *O. sativa,* where little effect of GA was reported, and Roberts (1963b), where some promotory effect of GA was reported on a cultivar of *O. sativa,* but full germination only occurred following further after-ripening after treatment with GA_3.

The effect of soaking in 2-mercaptoethanol solutions on imbibing seed

A number of compounds containing sulphydryl groups have been shown to break dormancy in seeds of cereal species (e.g. Pollock and Kirsop, 1956). In various rice seed lots thiourea (one such compound) has been reported as having a strong dormancy-breaking action if pre-applied to imbibed seeds at 0.5% for two hours (Murty and Raghavaiah, 1966); a moderate dormancy breaking action if co-applied at 0.2% (Nakamura, 1963); a very slight dormancy-breaking action when pre-applied or co-applied at 10^{-3}M–10^{-2}M (0.0076–0.076%) (Roberts, 1963b) or no effect when pre-applied or co-applied at 10^{-2}M–10^{-1}M (0.076–0.76%) (Roberts, 1963b). Since, overall, the action of thiourea did not seem particularly effective, we soaked seeds of the strongly dormant *O. glaberrima* cultivar G (table 1) for 24 hours in solutions of 2-mer-

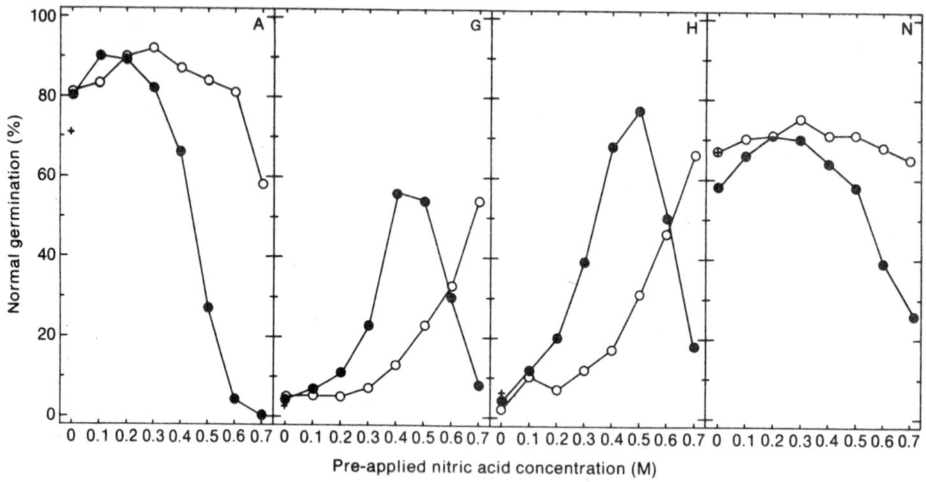

Figure 4. The effect of nitric acid pre-applied for either 12 (○) or 24 hours (●) (400 seeds in 50 ml) at various concentrations on the cumulative germination of 400 seeds of four rice cultivars (A, G, H, N) after 42 days in test at 20°/30°C between moist rolled paper towels and that of seed tested without pre-treatment (+).

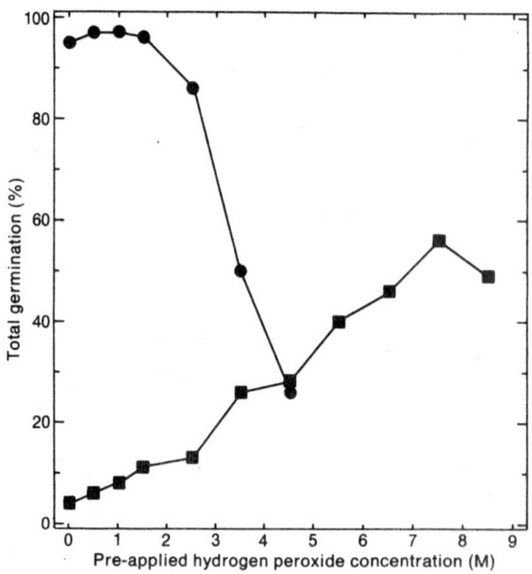

Figure 5. The effect of hydrogen peroxide pre-applied for one day (400 seeds in 200 ml) at various concentrations on the cumulative germination of 400 seeds of two rice cultivars (●, A, *O. sativa, indica*; ■, G, *O. glaberrima*) after 42 days in test at 20°/30°C between moist rolled paper towels.

captoethanol (400 seeds to 50 ml of solution), another sulphydryl compound which has been shown to have a particularly strong dormancy breaking action in barley at 0.05M (Roberts and Smith, 1977). The results of subsequent 400 seed germination tests in moist rolled paper towels at 20°/30 °C for 42 days showed no effect of 2-mercaptoethanol at concentrations of 0.01, 0.025, 0.05, 0.075 and 0.1M: in no case was germination greater than 6%. At first sight these results provide support to the conclusion that sulphydryl and related compounds show little or no effect on rice seed dormancy (Roberts, 1963b). However, in a later section we show that 2-mercaptoethanol can interact positively with other treatments.

The effect of potassium cyanide solutions on imbibing seed

Treatment with potassium cyanide has been shown to break seed dormancy in rice (Roberts, 1964a) and other cereals (Roberts and Smith, 1977). Seeds of the strongly dormant *O. glaberrima* cultivar G (table 1) were treated for 24 hours in 10^{-3}M KCN, but no difference was observed in subsequent germination over a 42 day test period at 20°/30 °C compared to a control (soaked in water for 24 hours) treatment. In view of this, and the fact that even if it were promotory, KCN is a poison whose use could only be recommended with extreme caution, no further investigations were carried out with this or similar compounds.

The effect of soaking in nitric acid on imbibing seed

Steeping dormant rice seeds in solutions containing either nitrate or hydrogen ions has been shown to promote germination (Roberts, 1963a). Moreover, steeping in a solution containing both nitrate ions and high concentrations of hydrogen ions simultaneously, that is in dilute nitric acid, has been shown to stimulate a number of dormant rice seeds to germinate which otherwise would have been unaffected by the nitrate or hydrogen ions alone (Roberts, 1963a). Steeping treatments of 12 hours in 0.01M HNO_3 (Roberts, 1963a), 12 hours in 0.1M HNO_3 (Jalote and Vaish, 1976), 16–20 hours in 0.1M HNO_3 (Subramoney and Abraham, 1969) and 23 hours in 0.1M HNO_3 (Agrawal and Nanda, 1969) have all been reported to stimulate the germination of otherwise dormant rice seed. Here (figure 4) seed of cultivars A, G, H and N were subjected to steeping treatments in dilute nitric acid over a range of concentrations and two treatment periods in an attempt to determine whether an optimum treatment level and period could be defined for four quite different cultivars (table 1).

The subsequent germination of all four cultivars provided evidence of both stimulation (increase) and inhibition (reduction), the effect depending upon the nitric acid concentration (figure 4). In two cultivars of *O. sativa*, A (*japonica*) and N (*javanica*), the stimulation was only slight because the majority of seeds were non-dormant (table 1), but in the strongly dormant cultivars of *O. glaberrima*, G and H, the maximum stimulation was substantial. There was a large reduction in germination following treatment at the highest nitric acid concentrations compared to lower concentrations in all four cultivars when soaked for 24 hours, and also in cultivars A and N only when soaked for 12 hours. Maximum germination was observed following soaking in different nitric acid concentrations for the four cultivars: after 24 hours soaking in 0.1M HNO_3 for cultivar A, 0.2M HNO_3 for cultivar N, 0.4M HNO_3 for cultivar G and 0.5M HNO_3 for cultivar H; and after 12 hours soaking in 0.3M HNO_3 for cultivars A and N and 0.7M HNO_3 for cultivars G and H. Within the range of treatment combinations investigated, soaking in dilute nitric acid was unable to promote the germination of all dormant seeds (see table 1) in cultivars G and H. Although soaking in dilute nitric acid has been advocated as a general procedure for promoting the germination of dormant rice seed (e.g. IRRI, 1968; Agrawal and Nanda, 1969), these results indicate that such treatments alone are insufficient for moderately or strongly dormant seed if the treatment concentration is limited to that which does not result in a loss in germination in weakly dormant rice seed lots.

The effect of soaking in hydrogen peroxide solutions on imbibing seed

Roberts (1964a) showed that hydrogen peroxide, pre-applied either at 8.5M for two hours or 1M for 24 hours, increased germination in a partially dormant rice seed lot from 40% to 85%. But four other oxidising agents (potassium chlorate, potassium dichromate, potassium permanganate and calcium hypochlorite) had no positive effect (Roberts, 1964a). Similarly Yasue (1973) has reported that treatment with 1% (0.3M) hydrogen peroxide increased rice seed germination.

For cultivar A, 0.5M and 1M H_2O_2 pre-applied for 24 hours resulted in maximum percentage germination (figure 5) to the level of viability indicated by a tetrazolium test (table 1), a very slight promotion of germination over the control treatment. However, treatment in higher concentrations of hydrogen peroxide resulted in a progressive reduction of percentage germination. For cultivar G the optimum treatment concentration for the promotion of germination of otherwise dormant seeds was 7.5M (figure 5), but not all the dormant seeds (table 1) were stimulated to germinate by this treatment. This concentration is very much greater than that (1M) found to be suitable for another rice seed lot by Roberts (1964a) and for cultivar A here (0.5M–1M). Thus we conclude that no single concentration of hydrogen peroxide was capable of promoting full germination of all lots and that high concentrations can sometimes reduce viability. Thus treatment in hydrogen peroxide alone is not a reliable practical technique.

The effect of constant and alternating temperatures on imbibed seed

In the preceding sections all tests for germination were carried out in a diurnal alternating temperature regime of 20°/30 °C, one of two regimes currently prescribed by ISTA (1976b, 1981). There is, however, little evidence that either this regime or the alternative (30 °C) provides conditions for the most regular, rapid and complete germination for the majority of rice seed lots, the object of the germination test (ISTA, 1976a).

A number of workers have investigated the responses of rice seed germination to different constant temperatures. With non-dormant seed full germination has been reported over quite wide constant temperature ranges, viz., 15°–37 °C (Livingston and Haasis, 1933), 26.5°–37.5 °C (Chaudhary and Ghildyal, 1969), 27°–42 °C (Roberts,

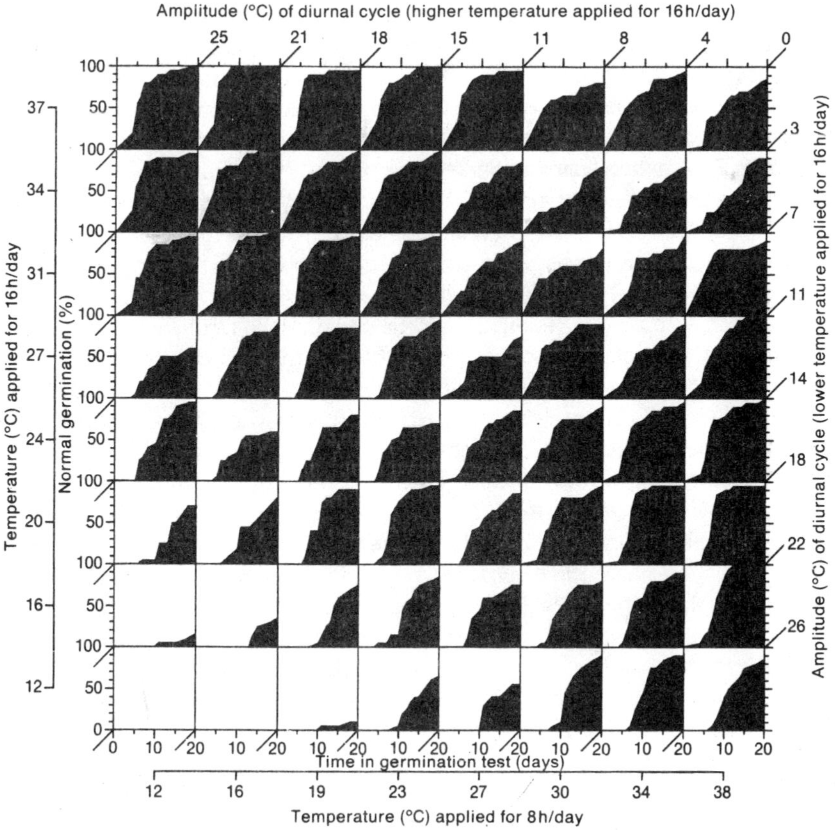

Figure 6. The influence of constant and alternating temperatures on the germination of seed of cultivar F (*O. sativa, japonica*) investigated on a two-dimensional temperature gradient plate over a period of 20 days. The plate was maintained on an 8 hour/16 hour cycle and provided 64 different temperature regimes. Twenty seeds were tested for germination in each of the 64 cells. Germination progress curves are presented for each environment and are superimposed on a plan of the plate to allow the temperature regime within each cell to be discerned. The area under each curve (blacked) is a combined function of the number of seeds which germinated and the time taken by each seed to germinate (but note the scales on the x axis in figures 9 and 10 are different from those in figures 6 to 8 and this needs to be noted when comparing the areas).

1962), 20°–30 °C (Nakamura, 1963; Ikehashi, 1975). Within these ranges germination is most rapid over much narrower ranges viz., 28°–36 °C (Akemine, 1914), 30°–35 °C (Livingston and Haasis, 1933), 27°–36 °C (Hall, 1966a, 1966b, 1966c), 32 °C (Chaudhary and Ghildyal, 1969), 30 °C (Ikehashi, 1975; Nishiyama, 1978). In partially dormant rice seed lots there is some evidence of induction of secondary dormancy at low temperatures, viz. 15 °C (Ikehashi, 1973, 1975). The optimum constant temperature for the germination of partially dormant rice seed occupies a very narrow band (above and below which percentage germination is reduced) and, apparently, differs between seed lots, viz., 27 °C (Roberts, 1962, 1965), 30 °C (Nakamura, 1963; Ikehashi, 1973, 1975), 32 °C (Hayashi and Morifuji, 1972). In addition it is possible that the optimum constant temperature may be dependent upon gaseous environment: in one seed lot percentage germination was higher at 27 °C than at 32 °C when tested in air, but higher at 32 °C than 27 °C when tested in nitrogen (Roberts, 1962); similarly germination and seedling growth were most rapid at 31 °C for aerated seed of two different cultivars, but most rapid at 33°–36 °C for seed submerged in water (Hall, 1966a); although in subsequent investigations the optimum temperature was 31 °C for seed in both environments (Hall, 1966b, 1966c).

There is considerably less information concerning the responses of rice seed germination to alternating temperatures. Moreover, what evidence there is is often based on studies limited to only a few or even one alternating temperature regime. Chaudhary and Ghildyal (1969), working with non-dormant seed, found no advantage of alternating temperature regimes over suitable constant temperatures. Nakamura (1963), working with partially dormant seed, found no difference in percentage germination between 20°/30 °C and 30 °C constant, but at 30°/15 °C percentage germination was greater than either of the above regimes. Hayashi and Morifuji (1972), working with partially dormant seed, found percentage germination at 20°/32 °C to be greater than at 32 °C constant, but both of these results were higher than at either 20°/26 °C or 26°/32 °C. Singh *et al* (1973) considered a constant temperature of 30 °C to be preferable to an alternating temperature regime of 20°/30 °C, but their results were equivocal since of the 20 rice seed lots tested, 10 gave the highest percentage germination when tested at 30 °C, eight at 20°/30 °C and two at 25 °C.

There is, therefore, some question as to whether constant or alternating temperatures provide the most suitable regime for the germination of both dormant and non-dormant rice seed lots. Moreover, whichever is the more suitable, the concept of an optimum temperature regime requires definition. We have attempted to answer these questions.

Concurrently with the preceding investigations, we investigated the germination responses of four rice seed lots to constant and alternating temperatures over a wide temperature range using a two-dimensional temperature gradient plate. The results are presented as a series of germination progress curves superimposed on plans of the temperature gradient plate (figures 6–10).

Two linear temperature gradients were imposed at right-angles to each other for 16 hours and eight hours respectively. The plate was divided into 64 squares and a treatment was allocated to each square. Across each square there was in fact a gradient of approximately 4 °C but there is little inaccuracy if each treatment (square) is considered as representing a single temperature-alternation treatment which may be taken as the two mean temperatures for each square: the temperature applied for 16 hours shown on the vertical axis and the eight hours treatment shown on the horizontal axis in figures 6–10. The mean temperatures for each period are shown to the nearest whole degree for each square and the two scales are not always exactly symmetrical.

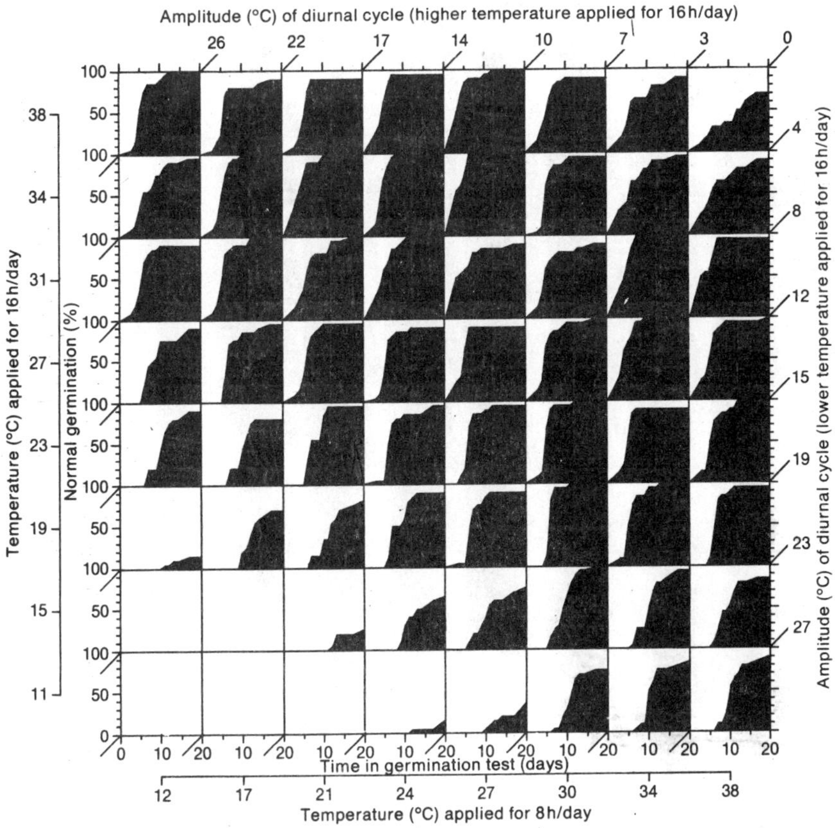

Figure 7. The influence of constant and alternating temperatures on the germination of seed of cultivar L (*O. sativa, indica*) investigated on a two-dimensional temperature gradient plate over a period of 20 days. Further explanation as for figure 6.

Nevertheless, it will be seen that the diagonal line of squares running from the bottom left to the top right of each diagram can be considered as a series of eight different constant temperatures since the alternation in any square does not exceed 4 °C and in fact averages less than this (figures 6–10).

Seed of four cultivars was used in this series of experiments. The first three investigations used seed of *O. sativa* which showed little dormancy; cultivars F (*japonica*) (figure 6), L (*indica*) (figure 7) and M (*javanica*) (figure 8). The remaining two investigations used seed of the deeply dormant *O. glaberrima* cultivar G, one of which was comparable to the above experiments (figure 9), and one in which the seeds were pre-treated with 0.5M HNO_3 at 20 °C for 12 hours (figure 10). It is clear that the most dormant cultivar (G) (figure 9) germinated more slowly in any given environment than the relatively non-dormant *O. sativa* cultivars (figures 6–8). Within the latter group, seed

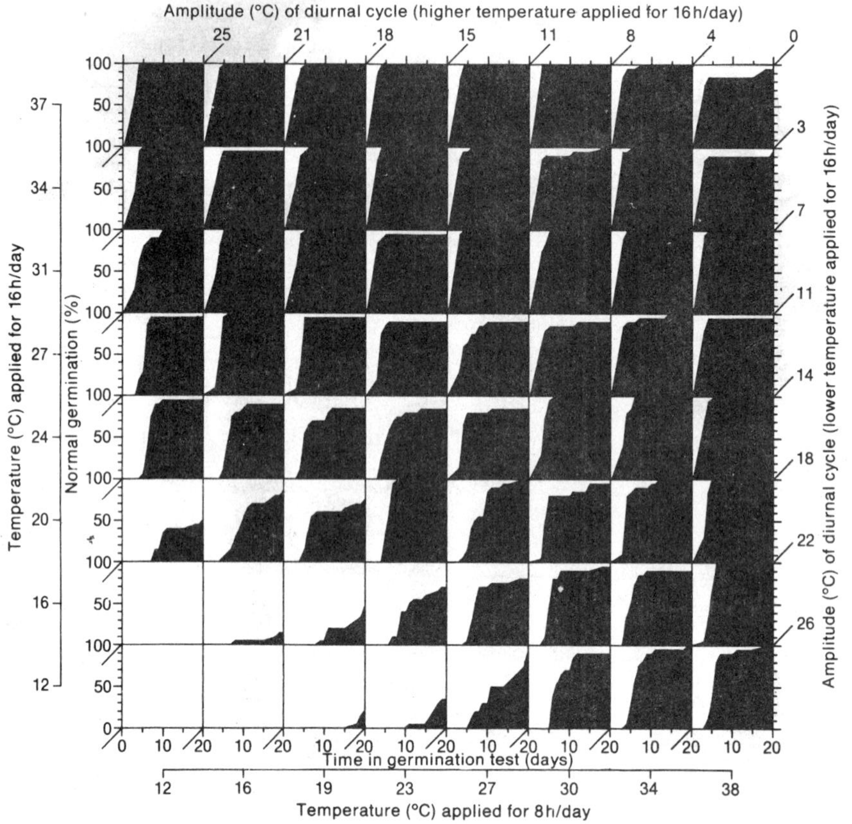

Figure 8. The influence of constant and alternating temperatures on the germination of seed of cultivar M (*O. sativa, javanica*) investigated on a two-dimensional temperature gradient plate over a period of 20 days. Further explanation as for figure 6.

of the *javanica* cultivar M (figure 8) germinated more rapidly than either the *japonica* cultivar F (figure 6) or the *indica* cultivar L (figure 7).

The extreme range of mean temperatures over which seeds of the *japonica* cultivar F showed at least some germination during the 20 day test period was 12°–38 °C (figure 6). No seed germinated in two cells, 12 °C and 12°/16 °C (mean 13 °C). Although some seed germinated at a constant temperature of 16 °C this was noticeably slower and less complete than at constant temperatures between 20°–38 °C. Within this range it was difficult to discern an optimum constant temperature either in terms of the rapidity of germination or the maximum germination achieved. However, at the higher constant temperatures 40%–50% of seeds germinated quite rapidly but the remaining seeds were somewhat delayed (figure 6); whereas at 20 °C the start of germination was delayed but the majority of seeds (90%) showed a tight distribution of germination peri-

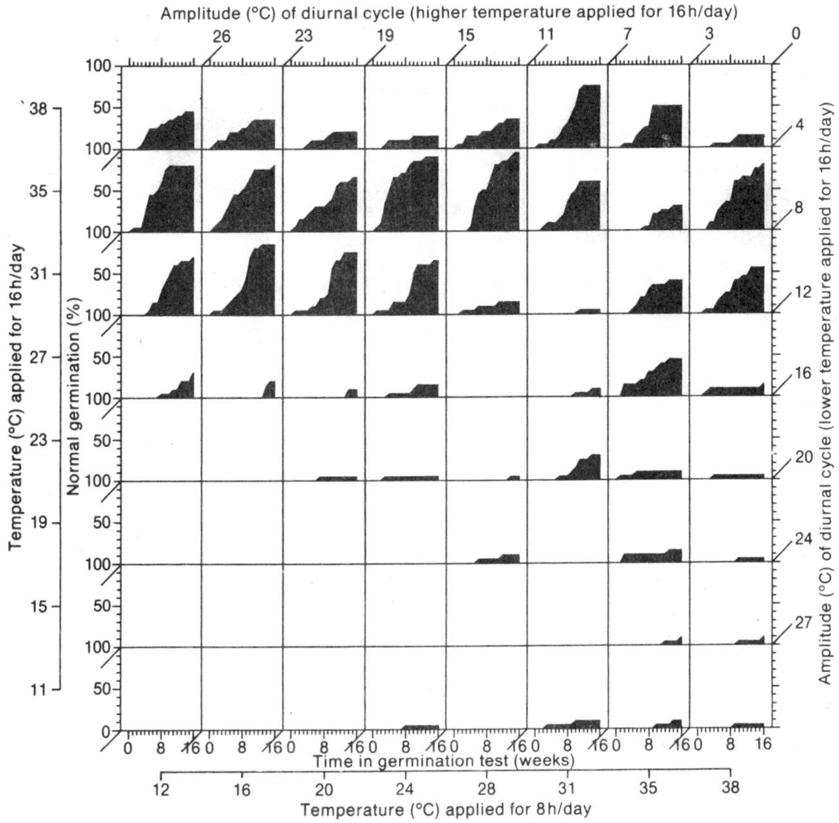

Figure 9. The influence of constant and alternating temperatures on the germination of seed of cultivar G (*O. glaberrima*) investigated on a two-dimensional temperature gradient plate over a period of 16 weeks. Further explanation as for figure 6.

ods, germinating between days 7 and 15. This, and the fact that this was the highest germination recorded at constant temperatures, suggests that 20 °C represents the optimum constant temperature for the germination of this lot. Germination was most rapid and complete (100%) in an alternating temperature regime of 37°/16 °C. Similar, though marginally less rapid, germination progress curves were obtained in other alternating temperature regimes, 16°–31°/34°, 21°–31°/38°, 31°/12°–23°, 34°/12°–23° and 37°/12°–27 °C. Ninety-five per cent germination was recorded at 20°/30 °C, the cell equivalent to the alternating temperature regime prescribed by ISTA, with a germination progress curve similar to that obtained at a constant temperature of 20 °C but twice as slow as the most suitable alternating temperature regime (37°/16 °C).

The extreme range of mean temperatures over which at least some seeds of the *indica* cultivar L germinated during a 20 day test was 15° to 38 °C, i.e. from an alternating

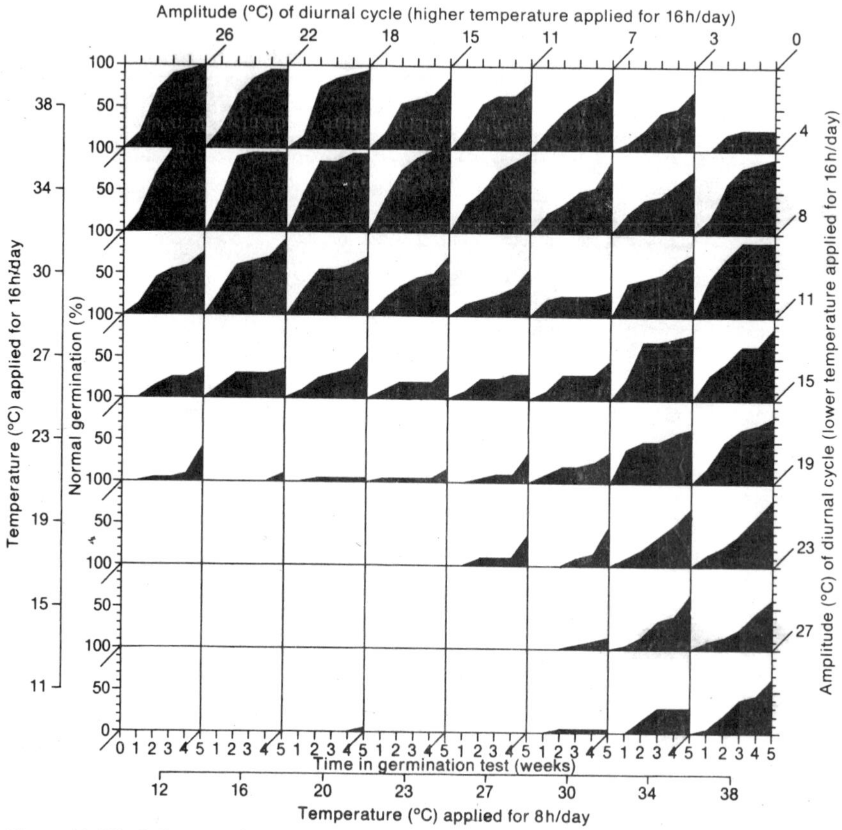

Figure 10. The influence of constant and alternating temperatures on the germination of seed of cultivar G (*O. glaberrima*) previously soaked for 12 hours in 0.5M nitric acid at 20 °C (1,280 seeds in 160 ml) investigated on a two-dimensional temperature gradient plate over a period of five weeks. Further explanation as for figure 6.

temperature regime of 11°/24 °C (mean 15 °C) to a constant temperature of 38 °C respectively (figure 7). Seeds were not tested for germination in environments with mean temperatures above 38 °C, but there was a clear cut off in germination at the lower temperature. Some seeds germinated at constant temperatures between 19°–38 °C but the totals failed to reach 100% during the 20 day test at all constant temperatures within this range. At constant temperatures germination occurred most rapidly at 27 °C, but the final germination at 20 days (90%) was similar to that at 23° (95%), 31° (90%) and 34 °C (95%). Results at constant temperatures of 16 °C and 20 °C provided a sharp contrast, no germination being recorded in the first compared to 80% in the second regime. Many alternating temperature regimes promoted full, rapid germination: 100% germination was recorded at 15°–27°/30°, 19°/34°, 27°–31°/34°,

23°–27°/38°, 31°/17°–24°, 34°/17°–27°, 38°/12° and 38°/27°C. Within these alternating temperature regimes germination occurred most rapidly at 34°/27°C. Complete germination was also recorded in the alternating temperature regime (19°/30°C), the closest to that prescribed by ISTA; however in this case germination was some days slower than at both the optimum alternating temperature regime, 34°/27°C, and the optimum constant temperature regime, 27°C.

Compared to the other cultivars (figures 6, 7, 9, 10) the progress of germination of seed of the *javanica* cultivar M was relatively unaffected by temperature regime, except where the temperature for part of the day failed to reach 30°C (figure 8). The extreme range of mean temperatures over which at least some seeds germinated during the 20 day test was 14° to 37°C, from an alternating temperature regime of 12°/19°C (mean 14°C) to a constant temperature of 37°C respectively. At constant temperatures the most rapid, complete germination occurred at 34°C, and, within the range of environments tested, the progress of germination in alternating temperature regimes was no more rapid than in this treatment. Indeed the germination progress curves in all cells where a temperature of 31°C or higher was applied for 16 hours per day or more were very similar with three exceptions, 31°/12°C, 34°/38°C and 37°C. Under the alternating temperature regime prescribed by ISTA, 20°/30°C, seed germination was somewhat delayed.

Figure 9 shows that the extreme range of mean temperatures over which seed of lot G germinated was 15° to 38°C, from an alternating temperature regime of 11°/24°C (mean 15°C) to a constant temperature of 38°C respectively, but seeds were not tested in environments where mean temperature exceeded 38°C. No germination occurred in environments where mean temperature was below 15°C nor where maximum temperature for part of each day failed to reach 23°C. Some seeds germinated at constant temperatures of 23°, 31°, 35° and 38°C, the highest germination (30%) being observed at 35°C. Many alternating temperature regimes promoted higher final, and more rapid, germination than did a constant temperature of 35°C; in general the superior alternating temperature regimes were between 35°/12°–38°C, between 31°/12°–24°C, between 38°/31°–35°C and at 31°/38°C and 27°/35°C. Within these alternating temperature regimes, the most rapid and complete (80%–95%) germination was observed between 35°/12°–28°C. There was no germination in the alternating temperature regime (19°/31°C) closest to that prescribed by ISTA for *O. sativa* and little germination in any neighbouring cells excepting 23°/31°C where germination (30%) equalled that in the most suitable constant temperature.

Figure 10 shows that when seed lot G was pre-treated with 0.5M HNO_3 for 12 hours germination was more rapid and achieved higher maximum values. However the range of mean temperatures over which it germinated was similar to seeds which had not been pre-treated (figure 9). Similarly the optimum constant temperature was 34°C. Some, but by no means all, alternating temperature regimes provided environments which resulted in higher final percentage germination and more rapid germination than at 34°C. In general the superior alternating temperatures were between 34°/12°–27°C, between 38°/12°–30°C and between 23°–34°/38°C. Germination in environments similar to the alternating temperature regime prescribed by ISTA for *O. sativa,* i.e. 19°–23°/30 °C, was considerably lower and slower than that at either 34 °C or the more superior alternating temperatures noted above. For any pair of maximum and minimum temperatures, germination was greater and more rapid if the maximum temperature was applied for 16 hours, rather than eight hours, per day, provided the maximum temperature did not exceed 34 °C; this contrasts with the regime currently recommended by ISTA in which the higher temperature is applied for the shorter period.

Alternating temperatures were more clearly of benefit for cultivar G when not pre-treated (figure 9) than for seed which was less dormant due to a pre-treatment in dilute nitric acid (figure 10). Nevertheless, in both cases alternating temperatures resulted in greater and more rapid germination than the most suitable constant temperature regime.

It is evident from these results on the two-dimensional temperature gradient plate (figures 6–10) that alternating temperature regimes exist which are more suitable than either constant temperatures (including that, 30 °C, prescribed by ISTA) or the alternating temperature of 20°/30 °C currently prescribed by ISTA for rice seed germination. It is, however, more difficult to specify exactly which particular regime should be chosen for germination tests. We originally chose 34°/11 °C, primarily because of the rapid germination observed in a similar regime of 35°/12°C for seed of cultivar G when pre-treated in dilute nitric acid (figure 10) since, in our opinion, a suitable alternating temperature regime should not only lead to maximum germination, but maximum germination within a convenient period (compare figures 9 and 10). However, as more data were accumulated from these investigations on the two-dimensional temperature gradient plate we began to question whether such a wide temperature alternation is necessary. Consequently in the final investigation (see table 2) seeds were also tested at 34°/23 °C.

On the temperature gradient plate the rate of change in temperature from the upper to the lower temperature, and vice-versa, was rapid (being accomplished within 5–10 minutes) and did not vary during these investigations. Thus we have no direct evidence as to whether this attribute of alternating temperature regimes affects loss in dormancy. Nevertheless comparing results obtained from testing the seed on the two-dimensional temperature gradient plate with subsequent tests within alternating temperature incubators does provide indirect evidence that the rate of temperature change may influence loss in rice seed dormancy. Seed of cultivar G tested at 34°–35°/11°–12°C for 42 days without any treatment showed 25% germination on the two-dimensional temperature gradient plate (figure 9), only 5% in an alternating temperature incubator (figure 11), but 34% in a further test in an alternating temperature incubator (table 2). Also at variance were the results obtained for cultivar I when tested at 34°/11 °C for 42 days in alternating temperature incubators without any treatment; in figure 11 germination was only 5%, whereas it was 22% in figure 12 and 29% in table 2. Finally cultivar H showed 14% germination in these conditions in figure 11, but 47% in table 2. For the results reported in figure 11 the incubators' refrigeration units operated continuously. However, before starting the investigations reported in figure 12 and table 2 the incubators were modified to enable the refrigeration units to operate only during the period spent at the lower temperature. The time taken to cool from the upper (34 °C) to the lower (11 °C) temperature remained unchanged at about 45 minutes, but the time taken to warm from the lower to the upper temperature was reduced from 140 to 45 minutes as a result of this modification. Thus higher germination of the dormant cultivars was observed when the rate of warming was rapid (figures 9 and 12, table 2) than when it was slower (figure 11). Of course the time spent at 34 °C was also increased as a result of this modification (closer to the notional 16 hours period at the upper temperature) and this, rather than the rate of change in temperature, could have been the important factor. Accordingly it is not possible from these results to decide conclusively that rate of change is important, but it is evident that slight differences in the alternating temperature regime can produce different results, even when the maximum and minimum temperatures and the switching times are similar.

Clearly earlier sections have shown that the isolated application of various dormancy breaking agents alone is insufficient to promote full and rapid germination of strongly dormant seed when tested at 20°/30 °C. Although pre-treatment in 0.5M nitric acid with subsequent testing at 35°/12 °C, and certain other alternating temperature regimes, was suitable for cultivar G (figure 10), it has already been noted that the germination of other rice seed lots is reduced following pre-treatment in nitric acid at this concentration (figure 4). However, work in other species demonstrates that a combination of dormancy breaking agents can produce a dramatic promotory effect on germination when the single factors have no, or little, effect (e.g. Roberts and Benjamin, 1979; Vincent and Roberts, 1977, 1979). Thus we investigated the combined influence of various dormancy breaking agents at concentrations which, it was hoped, would not reduce the germination of less dormant seed.

The combined effects of soaking in nitric acid and hydrogen peroxide, with tests in two alternating temperature regimes

We began by combining soaking pre-treatments in nitric acid and hydrogen peroxide (though, for reasons of chemical stability, not simultaneously) since both broke dormancy in some of the dormant seed and were benign to non-dormant seed at low concentrations (figures 4 and 5). In addition the seeds were tested at 20°/30 °C as well as 34°/11 °C to confirm whether the latter regime was more beneficial in the combined treatment.

There was little difference between the germination progress curves of dry seed or seed soaked for 48 hours in water in either alternating temperature regime, or between these regimes for seed in either of these categories (figure 11); the largest difference was observed for dry seed of cultivar N, which gave 80% germination after 42 days when tested at 34°/11 °C but only 70% at 20°/30 °C. A higher proportion of seed of the three dormant *O. glaberrima* cultivars (G, H, I) which had been pre-treated in dilute nitric acid germinated at 34°/11 °C than at 20°/30 °C. The influence of dilute

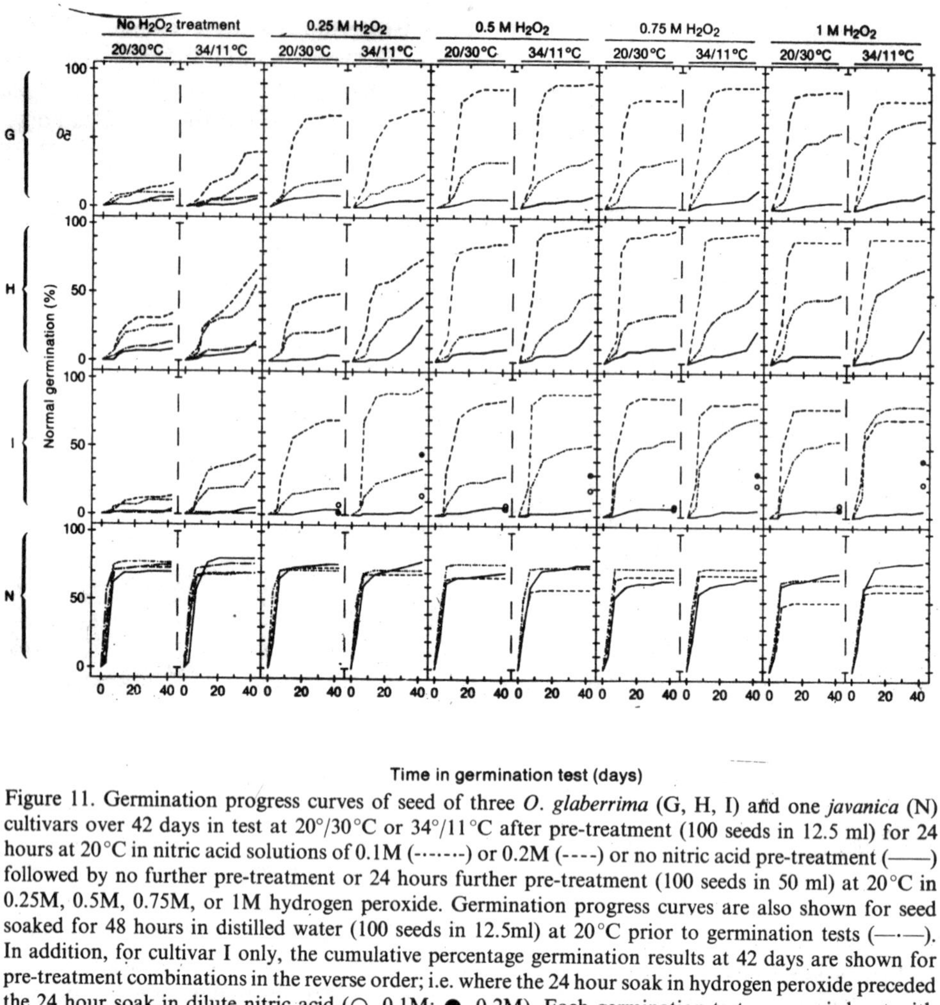

Figure 11. Germination progress curves of seed of three *O. glaberrima* (G, H, I) and one *javanica* (N) cultivars over 42 days in test at 20°/30°C or 34°/11°C after pre-treatment (100 seeds in 12.5 ml) for 24 hours at 20°C in nitric acid solutions of 0.1M (-·-·-·-) or 0.2M (----) or no nitric acid pre-treatment (——) followed by no further pre-treatment or 24 hours further pre-treatment (100 seeds in 50 ml) at 20°C in 0.25M, 0.5M, 0.75M, or 1M hydrogen peroxide. Germination progress curves are also shown for seed soaked for 48 hours in distilled water (100 seeds in 12.5ml) at 20°C prior to germination tests (—·—). In addition, for cultivar I only, the cumulative percentage germination results at 42 days are shown for pre-treatment combinations in the reverse order; i.e. where the 24 hour soak in hydrogen peroxide preceded the 24 hour soak in dilute nitric acid (○, 0.1M; ●, 0.2M). Each germination test was carried out with 100–110 seeds divided amongst four 9 cm petri dishes and germinated on filter paper wetted with 4.5 ml distilled water.

nitric acid alone on these three cultivars at these concentrations was small but discernible, more especially in tests at 34°/11°C. In all cases germination following pre-treatment in 0.2M HNO_3 was greater than for seed treated in 0.1M HNO_3 and each was greater than the control treatments. The form of the germination progress curves at 34°/11°C were distinct from those at 20°/30°C for all seed of cultivars G, H, I showing intermediate levels of germination (10% to 70%) in these tests; at 20°/30 °C little further germination occurred after 21 days in test, whereas at 34°/11 °C germination continued to occur over the subsequent 21 day test period. This is clear evidence that the alternating temperature regime of 34°/11 °C was, apart from any influence of dilute nitric acid or hydrogen peroxide either singly or in combination, promoting the germination of at least some of the seeds which remained dormant in the 20°/30 °C regime. The influence of dilute nitric acid alone on the germination of seed of cultivar N was less marked. At 20°/30 °C germination was slightly higher, but at 34°/11 °C slightly lower than for control treatments. However, in both cases the differences are within the range of variation to be expected from 100 seed germination tests. Hydrogen peroxide at all concentrations pre-applied on its own had no effect on promoting the germination of dormant seed tested at 20°/30 °C, and only a small effect in cultivars G and H when tested at 34°/11 °C. Even where the influence of hydrogen peroxide on its own was observed, dormancy was only lost in some seed after 21 to 28 days exposure of the imbibed seed to 34°/11 °C.

Some, but not all, combinations of pre-treatments in dilute nitric acid and hydrogen peroxide provided evidence of a positive first-order interaction in the promotory effect of the two agents; e.g. for seed of cultivars G and I when tested at 34°/11 °C following a 24 hour pre-treatment in 0.2M HNO_3 and a subsequent 24 hours pre-treatment in 0.25M H_2O_2 (figure 11). It is clear that this sequence of application is essential to obtain the combined promotory effect, for in the reverse order the cumulative percentage germination of cultivar I after 42 days in test at 20°/30 °C was no more than control test results. Germination was higher than control tests at 34°/11 °C but no more than the appropriate result following treatment in dilute nitric acid alone. For all 16 combinations germination was substantially less than where treatment in dilute nitric acid preceded the treatment in hydrogen peroxide.

There was a possibility that the difference observed when the order of application was reversed (figure 11) was not due to the order of application of the two agents, but because hydrogen peroxide promotes germination much more if applied to imbibed seed. To test this hypothesis seeds of cultivar I were treated with hydrogen peroxide after soaking in water. The results of 42 day germination tests at 20°/30 °C for a control treatment (48 hour soak in water, 2%), and 24 hour soak in water followed by a further 24 hour soak in 0.25M hydrogen peroxide (6%), 0.5M hydrogen peroxide (8%), 0.75M hydrogen peroxide (9%) or 1M hydrogen peroxide (10%) confirmed that the promotion of germination recorded after consecutive soaking treatments in dilute nitric acid and hydrogen peroxide (cultivars G, H, I, figure 11) did require the application of both agents in this order and was not the result of hydrogen peroxide alone acting on previously imbibed seed.

Maximum germination for cultivars G (91%) and H (98%) was observed when tested at 34°/11 °C following pre-treatment in 0.2M HNO_3 and 0.5M H_2O_2, for cultivar I (93%) when tested at 34°/11 °C following pre-treatment in 0.2M HNO_3 and 0.25M H_2O_2, and for cultivar N (80%) when tested at 34°/11 °C without pre-treatment (figure 11). For cultivar N the results of germination tests following treatment in both dilute

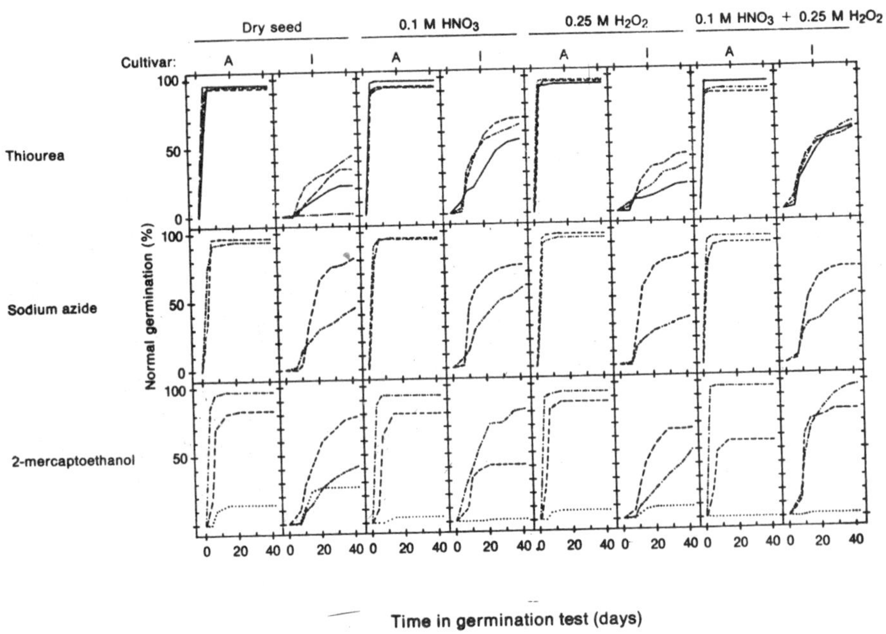

Figure 12. Germination progress curves of seed of two rice cultivars (A, I) in test at 34°/11 °C for 42 days in 9 cm petri dishes on filter papers wetted with distilled water (——), 0.02M (-·-·-·-) or 0.05M (----) thiourea, 0,0001M (-·-·-·-) or 0.001M (----) sodium azide, 0.01M (-·-·-·-) 0.025M (----) or 0.05M (........) 2-mercaptoethanol. Prior to these tests the seeds were subjected to no pre-treatment, soaked for 24 hours in 0.1M HNO_3 or 0.25M H_2O_2, or soaked in 0.1M HNO_3 for 24 hours and then in 0.25M H_2O_2 for a further 24 hours. In addition as a further treatment seeds were soaked for 24 hours in distilled water (—·—). All soaking treatments were conducted at 20 °C with 100 seeds to 12.5 ml of dilute nitric acid or water, or 50 ml of hydrogen peroxide. Each curve is the result of a germination test with 100–110 seeds.

nitric acid and hydrogen peroxide were lower than control tests and quite variable. However, only in a few treatments (e.g. 0.2M HNO_3, 1M H_2O_2) was there a clear reduction in germination following pre-treatment.

If a single optimum treatment were to be chosen as that which promoted the most germination of seed of the dormant cultivars whilst not reducing the germination of the relatively non-dormant cultivar N greatly, then the results presented in figure 11 suggest a 24 hour soak in 0.2M HNO_3 followed by a 24 hour soak in 0.25M, or possibly 0.5M H_2O_2 and germination at 34°/11 °C. Clearly these treatments, particularly the latter, are effective in promoting the germination of very dormant seed. However, we were concerned whether such a treatment would be too harsh for seed lots with only a small proportion of dormant seed. Consequently, seeds of cultivar A were tested at 34°/11 °C following no pre-treatment, or a 48 hour soak in water, or a 24 hour soak in 0.2M HNO_3 followed by 24 hours in 0.25M H_2O_2, or 24 hour soak in 0.2M

HNO_3 followed by 24 hours in 0.5M H_2O_2. Unfortunately the results after 42 days in test (95%, 92%, 82% and 46% respectively) demonstrated that the combined effect of nitric acid and hydrogen peroxide at these concentrations was deleterious for such seed.

Dilute nitric acid and hydrogen peroxide, when combined, strongly promote the germination of dormant seed (figure 11). Indeed, comparison with figures 4 and 5 demonstrates that far more dilute concentrations are adequate if both dormancy-breaking agents are applied. Nevertheless, despite the application of these agents in the more dilute form, there is still evidence that treatment at the concentrations required to promote germination of very dormant seed harm a proportion of non-dormant seed. Thus combining nitric acid and hydrogen peroxide soaking treatments at these concentrations is not an adequate, standard, dormancy-breaking procedure.

The combined effect of pre-applied nitric acid and hydrogen peroxide with co-applied thiourea, sodium azide or 2-mercaptoethanol

Although the germination of cultivar A was reduced following treatment in 0.2M HNO_3 and 0.25M H_2O_2, the effect was small. We suspected that more dilute concentrations might be more appropriate since each agent alone at low concentrations was not injurious (figures 4 and 5). However, such a treatment would be insufficient to promote the rapid germination of the more dormant seed. Thus an additional dormancy-breaking agent was required.

Thiourea, sodium azide or 2-mercaptoethanol were co-applied as the possible additional agent, despite previous results where pre-applied 2-mercaptoethanol alone failed to promote the germination of cultivar G when subsequently tested at 20°/30°C. (In preliminary experiments, not reported, a considerable number of other agents were tested in combination with nitric acid and hydrogen peroxide. Of these gibberellic acid was quite effective in further promoting germination, but the subsequent development of seedlings was abnormal with extended coleoptiles and stunted roots.)

As postulated (above) the germination of seed of cultivar A was not reduced, compared to a control, following 24 hours soak in 0.1M HNO_3 and a subsequent 24 hour treatment in 0.25M H_2O_2 (figure 12). Neither was germination reduced if non pre-treated seeds were tested with co-applied 0.02M or 0.05M thiourea, 0.0001M or 0.001M sodium azide, or 0.01M 2-mercaptoethanol. However, higher concentrations of 2-mercaptoethanol (0.025M or 0.05M) were injurious to these seeds both with and without pre-treatment. Co-applied thiourea and 0.001M sodium azide were slightly injurious to these seeds when pre-treated in both 0.1M HNO_3 and 0.25M H_2O_2.

Pre-treatment of seed of cultivar I with 0.1M HNO_3 followed by 0.25M H_2O_2 promoted germination from 22% to 60% but failed to break the dormancy of all seeds (figure 12, table 1). Thiourea co-applied to seed not pre-treated or to seed pre-treated with either 0.1M HNO_3 or 0.25M H_2O_2 promoted germination slightly, but no further promotion was observed when thiourea was co-applied to seed pre-treated in both 0.1M HNO_3 and, subsequently, 0.25M H_2O_2 (figure 12). Co-applied sodium azide also promoted germination: at 0.001M promotion occurred with or without pre-treatment to the seed; at 0.0001M the promotion was only slight for dry seed or seed pre-treated with either 0.1M HNO_3 or 0.25M H_2O_2; but no effect of 0.01M sodium azide was observed on seed pre-treated with both nitric acid and hydrogen peroxide. Co-applied 0.05M 2-mercaptoethanol was injurious to the seed of cultivar I, save for that not pre-treated where there was a possible slight promotion. Co-applied 0.01M or 0.025M 2-mercaptoethanol promoted germination of dormant seed in all cases save one where the seeds were tested with 0.025M 2-mercaptoethanol following treatment in 0.1M HNO_3. Maximum germination (95%) of cultivar I was observed following 24 hours soak in 0.1M HNO_3, a further 24 hours soak in 0.25M H_2O_2 and testing with co-applied 0.01M 2-mercaptoethanol. Moreover this treatment did not injure the germination (95%) of cultivar A. This suggested that these conditions might prove to be a suitable, standard, dormancy-breaking procedure.

To test this possibility seed of all 14 cultivars (table 1) was each tested under six separate regimes. These factorially combined either sequential soaking in 0.1M HNO_3 and 0.25M H_2O_2 and testing with co-applied 0.01M 2-mercaptoethanol or dry seed tested in water with three different alternating temperature environments: 20°/30°C, 34°/23°C or 34°/11°C. Three temperature regimes were investigated to allow comparison of the results with ISTA prescribed test conditions and to determine whether the broad amplitude of temperature alternation used in this and the previous section is necessary (see figures 6–10).

Analysis of variance showed significant effects ($P < 0.005$) of all three factors (cultivar, temperature, soaking) their first order interactions (cultivar × temperature, cultivar × soaking, soaking × temperature) and the second order interaction (cultivar × temperature × soaking). Of the six treatments investigated, no single treatment was optimal for the germination of seed of all 14 cultivars (table 2). Soaking in 0.1M HNO_3 for 24 hours followed by a further 24 hours soaking in 0.25M H_2O_2 and testing on filter paper moistened with 0.01M 2-mercaptoethanol at 34°/11°C provided optimal germination conditions resulting in either the highest germination or a value not significantly less than the maximum value recorded, for 10 of the 14 cultivars (A, B, C, E, G, H, I, L, M, N). The ISTA prescribed test procedure, dry seed tested at 20°/30°C, provided optimal germination conditions for nine of the 14 cultivars (A, B, C, E, F, J, K, L, M).

The germination of the least dormant seed lots of *O. sativa*, i.e. *japonica* cultivars B, F, J and K, *javanica* cultivar M, and *indica* cultivars A and L (table 1), was high in all treatments and in general there were few differences between treatments. Noteworthy exceptions were: a slightly inhibitory effect of the soaking treatment in some cultivars (F, J, K, M); the germination of dry seed of cultivar J at 34°/11°C was significantly lower than at either 20°/30°C or 34°/23°C; and the germination of dry seed of cultivar K at 20°/30°C was significantly higher than at either 34°/23°C or 34°/11°C. In the more dormant *javanica* cultivar (N), pre-treatment was an advantage when the seeds were tested at 20°/30°C and brought germination up to the level attained in the other two alternating temperature treatments without pre-treatment (table 2) and to the level of viability indicated in a tetrazolium test (table 1). In the more dormant *indica* cultivars the best (D) or equal best (E) results were obtained at 34°/11 °C without pre-treatment. For all the *O. glaberrima* cultivars, whether non-dormant (C) or particularly when deeply dormant (G, H, I), the highest germination was recorded when the soaking treatments were combined with subsequent testing at 34°/11 °C with co-applied 2-mercaptoethanol.

Concluding remarks

It had been our aim to develop a germination test procedure that could be applied to all rice seed lots, whether dormant or not, which would promote the germination of all viable seeds. This we have been unable to do. However, we believe that it is possible to make suggestions for procedures that are suitable for each of the taxonomic groups of rice.

Where considerable dormancy is expected, i.e. in *O. glaberrima* or in some *indica* cultivars of *O. sativa,* particularly when germinated soon after harvest, the procedure developed here, viz. soak at 20 °C for 24 hours in 0.1M HNO_3, then soak for a further 24 hours in 0.25M H_2O_2 followed by germination in 0.01M 2-mercaptoethanol in an alternating temperature regime of 16 hours at 34 °C and eight hours at 11 °C, is a better and more widely applicable prescription for removing dormancy than treatments suggested previously. Not only does this procedure remove dormancy effectively from very dormant seed and lead to rapid and normal germination but, unlike some other recommended treatments, it does not cause damage to less dormant *O. glaberrima* samples. However, it is possible that less dormant seed of some *indica* samples may be damaged by the soaking treatment. In this case the above alternating temperature regime (34°/11 °C) alone (that is without the chemical treatments) would appear to be a more suitable germination procedure than either the combined treatment described above or the alternating temperature regime (20°/30 °C) prescribed by ISTA.

The existing ISTA prescription (germination in an alternating temperature regime of 16 hours at 20°C and eight hours at 30 °C) appears to be most suitable for those samples of *japonica* cultivars which show little dormancy. However, the use of the ISTA recommended method of removing dormancy (pre-soak for 24–48 hours in water at 40 °C) is questioned as not being very effective and, more important, being potentially damaging to some cultivars. Only two *javanica* cultivars were included in this investigation and these were only slightly dormant. The results suggest, however, that the alternating temperature regime 34°/11 °C is preferable to the ISTA prescription (because the former regime is more widely applicable) and, in this regime, there was no advantage in incorporating the chemical treatments. Both *japonica* and *javanica* cultivars typically show little dormancy. Nevertheless, if pronounced dormancy were encountered in samples of these ecogeographical races it would be worthwhile trying the combined procedure developed here, particularly as no significant damage to germination was caused in any of the six samples of *japonica* and *javanica* cultivars used in this investigation by the combined treatment at 34°/11 °C compared with the result at 34°/11 °C without the chemical treatments (table 2).

Table 2. Percentage germination (with means of arcsin transformed data in parenthesis) of 14 rice cultivars after 42 days in three alternating temperature regimes (400 seeds per test), either for dry seed tested on filter paper wetted with distilled water, or for seed tested on filter paper wetted with 0.01M 2-mercaptoethanol after a 24 hour soak in 0.1M HNO_3 (400 seeds in 50 ml) and a further 24 hour soak in 0.25 M H_2O_2 (400 seeds in 200 ml).

Cultivar	Temperature (°C)					
	20/30		34/23		34/11	
	dry	soak	dry	soak	dry	soak
A	94	92	92	91	95	93
	(76.3)	(74.2)	(73.3)	(73.0)	(77.2)	(74.4)
B	93	93	93	90	93	90
	(74.9)	(74.4)	(74.8)	(71.3)	(74.6)	(71.4)
C	87	87	81	86	79	89
	(69.1)	(69.2)	(64.5)	(68.4)	(63.0)	(70.9)
D	84	87	92	83	97	85
	(66.4)	(69.0)	(74.0)	(65.7)	(80.3)	(67.2)
E	68	62	68	66	70	73
	(55.6)	(52.2)	(55.6)	(54.1)	(57.0)	(58.8)
F	97	95	97	93	97	95
	(82.4)	(77.1)	(80.7)	(74.3)	(80.3)	(76.7)
G	5	18	11	72	34	85
	(12.1)	(25.3)	(19.5)	(58.3)	(35.6)	(67.6)
H	4	20	30	80	47	92
	(12.0)	(26.5)	(33.0)	(63.5)	(43.5)	(74.2)
I	2	19	10	82	29	89
	(8.5)	(25.6)	(18.3)	(65.2)	(32.4)	(70.6)
J	98	93	99	93	95	93
	(82.1)	(74.8)	(83.5)	(75.5)	(77.6)	(75.0)
K	99	95	96	96	96	95
	(85.4)	(77.9)	(79.0)	(79.1)	(78.1)	(76.7)
L	97	96	99	96	97	96
	(80.5)	(78.5)	(83.5)	(78.7)	(80.5)	(79.1)
M	96	96	98	95	99	96
	(79.1)	(78.5)	(82.3)	(76.7)	(83.5)	(79.0)
N	72	80	78	80	81	78
	(58.3)	(63.7)	(62.0)	(63.8)	(64.2)	(62.1)

l.s.d. (0.05) = 5.0 (arcsin transformed data). For accuracy arcsin transformed data have been reported to the nearest 0.1%, whereas raw germination data have been reported to the nearest 1% for clarity. Consequently some treatment results have the same raw value, but their arcsin transformed values differ.

These results suggest that it may be worth re-examining the ISTA Rules for *O. sativa,* possibly making a distinction between the three taxonomic groups. We realise that the procedure suggested here for *indica* and *javanica* cultivars involves an alternating temperature regime which may be impossible for all seed testing laboratories to achieve at present. Nevertheless it seems reasonably clear that, whichever of the two alternating temperature regimes is used, it is preferable for all three ecogeographical races of *O. sativa* (and for *O. glaberrima*) that the higher temperature be applied for the longer period in the 24 hour cycle rather than the shorter period as is currently prescribed.

The germination progress curves presented here confirm that the 14 day test period prescribed by ISTA for *O. sativa* is, in general, adequate, especially if the rules were modified so that the higher temperature is applied for the greater part of the 24 hour cycle. At present, because of its limited distribution, there are no rules for *O. glaberrima;* but the results described here suggest that 14 days is an inadequate test period for this species, even when the recommended dormancy-breaking procedure we have developed is used.

For breaking dormancy of large bulks of seeds in breeding programmes the best technique would seem to depend on the taxonomic affiliation. IRRI (1968) has reported that pre-soaking in 0.1M HNO_3 for 16–24 hours followed by three to seven days sun-drying is suitable for the majority of weakly and moderately dormant seed lots. It might be worth investigating the possibility of following the HNO_3 treatment with 0.25M H_2O_2 before sun-drying for dealing with more dormant seeds. An alternative is to after-ripen dry seeds (about 11% moisture content) at about 47 °C for about a week. This treatment should have little effect on viability (Roberts, 1961c) and is probably suitable for most genotypes of *O. sativa.* However it is clear that such a treatment is not practicable for the normally deeply dormant genotypes of *O. glaberrima* since the period required at 47 °C is longer and consequently leads to loss of viability, and at a lower and safer temperature the treatment period required is too long to be of practical value. Thus in this species it is suggested that treatment with HNO_3, H_2O_2 and mercaptoethanol may be the best approach currently available; but the success may depend on the extent of the temperature conditions experienced in the field, and it might be wise to sow the seeds as near to the soil surface as possible in order to expose them to a wide temperature alternation.

Dormancy-breaking treatments are needed for use in the seed laboratories associated with seed banks for genetic conservation. In this context it is particularly important to use good quality seed lots to ensure maximum longevity (Ellis and Roberts, 1980), and this means getting the accessions into store as soon as possible after harvest. Thus dormancy is likely to be a considerable problem when using germination tests to assess initial viability, and indeed in subsequent viability monitoring tests during storage since dormancy (as well as viability) is preserved at low temperatures. We believe that the procedures described here will provide a suitable basis for such tests.

Acknowledgements

We are most grateful to Dr. T. T. Chang of the International Rice Research Institute and Mr. G. Banya of the Rice Research Station, Sierra Leone for supplies of seed, and to the International Board for Plant Genetic Resources for financial support.

References

Agrawal, P. K. and Nanda, J. S. (1969). A note on dormancy in rice. *Riso,* **18**, 325–326.

Akemine, M. (1914). Zur Kenntnis der Keimungsphysiologie von *Oryza sativa* (Reis). *Fühlings Landwirtschaftlich Zeitung,* **63**, 78–93. (Cited and re-analysed by Livingston and Haasis, 1933.)

Baskin, J. M. and Baskin, C. C. (1979). Effect of relative humidity on after-ripening and viability in the seeds of the winter annual *Draba verna. Botanical Gazette,* **140**, 284–287.

Carpenter, A. J. and Roberts, E. H. (1962). Some useful techniques in speeding up rice breeding programmes. *Empire Journal of Experimental Agriculture,* **30**, 127–131.

Chang, T. T. (1976a). Characterization and maintenance of genetic stocks. In *Manual on genetic conservation of rice germ plasm for evaluation and utilisation* (ed. T. T. Chang), pp. 12–28. International Rice Research Institute, Los Baños, The Philippines.

Chang, T. T. (1976b). The rice cultures. *Philosophical Transactions of the Royal Society of London, Series B,* **275**, 143–157.

Chaudhary, T. N. and Ghildyal, B. P. (1969). Germination response of rice seeds to constant and alternating temperatures. *Agronomy Journal,* **61**, 328–330.

Delouche, J. C. and Nguyen, N. T. (1964). Methods for overcoming seed dormancy in rice. *Proceedings of the Association of Official Seed Analysts of North America,* **54**, 41–49.

Dore, J. (1955). Dormancy and viability of paddy seed. *Malayan Agricultural Journal,* **38**, 163–173.

Ellis, R. H., Hong, T. D. and Roberts, E. H. (1982). An investigation of the influence of constant and alternating temperature on the germination of cassava seed using a two-dimensional temperature gradient plate. *Annals of Botany,* **49**, 241–246.

Ellis, R. H. and Roberts, E. H. (1980). Improved equations for the prediction of seed longevity. *Annals of Botany,* **45**, 13–30.

Ellis, R. H. and Roberts, E. H. (1981). The quantification of ageing and survival in orthodox seeds. *Seed Science and Technology,* **9**, 373–409.

Ghosh, B. N. (1960). Studies on germination of spring (boro) and summer (aus) paddy to determine dormancy and viability. *Indian Journal of Agricultural Science,* **30**, 129–134.

Hall, V. L. (1966a). Temperature and the germinating rice seed. I. Minimum to maximum temperature for growth in four days. *Rice Journal,* **69**, 40–42.

Hall, V. L. (1966b). Temperature and the germinating rice seed. II. Effect of temperature on germination and eight days growth of aerated seed. *Rice Journal,* **69**, 22–23.

Hall, V. L. (1966c). Temperature and the germinating rice seed. III. Effect of temperature on germination and eight days growth of submerged seeds. *Rice Journal,* **69**, 14–15.

Hayashi, M. (1977). [Studies on dormancy and germination of rice seed. V. The effects of the high temperature treatment on breaking the dormancy of the high moisture seeds in rice.] *Japanese Journal of Tropical Agriculture,* **22**, 164–171.

Hayashi, M. (1980). [Studies on dormancy and germination of rice seed. IX. The effects of oxygen and moisture upon the release of the rice seed dormancy and upon the inactivation of inhibitors in the dormant seed.] *Bulletin of the Faculty of Agriculture, Kagoshima, University,* **30**, 1–9.

Hayashi, M. and Morifuji, N. (1972). [Studies on the dormancy and germination of rice seed. I. The influences of temperatures and gaseous conditions on dormancy and germination in rice seeds.] *Japanese Journal of Tropical Agriculture,* **16**, 115–120.

Ikeda, M. (1963). [Studies on the viviparous germination of rice seed.] *Bulletin of the Faculty of Agriculture, Kagoshima, University,* **13**, 89–115.

Ikehashi, H. (1970). Germination habits in rice breeding. *Japanese Agricultural Research Quarterly,* **4**, 1–4.

Ikehashi, H. (1973). [Studies on the environmental and varietal differences of germination habits in rice seeds with special reference to plant breeding.] *Journal of the Central Agricultural Experiment Station, Konosu,* **19**, 1–60.

Ikehashi, H. (1975). Dormancy formation and subsequent changes of germination habits in rice seeds. *Japanese Agricultural Research Quarterly,* **9**, 8–12.

International Rice Research Institute (1968). Seed dormancy. *IRRI Reporter,* **4**, 1–4.

International Rice Research Institute (1972). *Annual Report for 1971*, pp. 20–21. International Rice Research Institute, Los Baños, The Philippines.

International Rice Research Institute (1973). *Annual Report for 1972*, pp. 18–19. International Rice Research Institute, Los Baños, The Philippines.

International Seed Testing Association (1976a). International rules for seed testing. Rules 1976. *Seed Science and Technology*, **4**, 3–49.

International Seed Testing Association (1976b). International rules for seed testing. Annexes 1976. *Seed Science and Technology*, **4**, 51–177.

International Seed Testing Association (1981). *Amendments to International Rules for Seed Testing 1976*. International Seed Testing Association, Zurich.

Jalote, S. R. and Vaish, C. P. (1976). Dormancy behaviour of paddy varieties in U.P. *Seed Research*, **4**, 187–190.

Jennings, P. R. and de Jesus, J. (1964). Effect of heat on breaking seed dormancy in rice. *Crop Science*, **4**, 530–533.

Livingston, B. E. and Haasis, F. W. (1933). Relations of time and maintained temperature to germination percentage for a lot of rice seed. *American Journal of Botany*, **20**, 596–615.

Mikkelsen, D. S. and Sinah, M. N. (1961). Germination inhibition in *Oryza sativa* and control by preplanting soaking treatments. *Crop Science*, **1**, 332–335.

Misra, P. K. and Misro, B. (1970). Seed dormancy in the African cultivated rice (*Oryza glaberrima* Steud.). *Indian Journal of Agricultural Science*, **40**, 13–16.

Murty, K. S. and Raghavaiah, P. (1966). Observations on dormancy in rice seed. *Current Science*, **35**, 548.

Nakamura, S. (1963). Short communication on dormancy of rice seed. *Proceedings of the International Seed Testing Association*, **28**, 57–59.

Nishiyama, I. (1978). Further evidence for the break on the Arrhenius plot of germination activity in rice seeds. *Japanese Journal of Crop Science*, **47**, 557–562.

Oka, H. I. and Tsai, K. H. (1955). [Dormancy and longevity of rice seed with regard to their variations among varieties.] *Japanese Journal of Breeding*, **5**, 22–26.

Pollock, J. R. A. and Kirsop, B. H. (1956). Studies in barley and malt. VI. Stimulation of the germination of freshly-harvested barley. *Journal of the Institute of Brewing*, **62**, 323–327.

Quail, P. H. and Carter, O. G. (1969). Dormancy in seeds of *Avena ludoviciana* and *A. fatua*. *Australian Journal of Agricultural Research*, **20**, 1–11.

Roberts, E. H. (1961a). Dormancy of rice seed. I. The distribution of dormancy periods. *Journal of Experimental Botany*, **12**, 319–329.

Roberts, E. H. (1961b). Dormancy in rice seed. II. The influence of covering structures. *Journal of Experimental Botany*, **12**, 430–445.

Roberts, E. H. (1961c). The viability of rice seed in relation to temperature, moisture content and gaseous environment. *Annals of Botany*, **25**, 381–390.

Roberts, E. H. (1962). Dormancy in rice seed. III. The influence of temperature, moisture and gaseous environment. *Journal of Experimental Botany*, **13**, 75–94.

Roberts, E. H. (1963a). The effects of inorganic ions on dormancy in rice seed. *Physiologia Plantarum*, **16**, 732–744.

Roberts, E. H. (1963b). The effects of some organic growth substances and organic nutrients on dormancy in rice seed. *Physiologia Plantarum*, **16**, 745–755.

Roberts, E. H. (1963c). An investigation of inter-varietal differences in dormancy and viability of rice seed. *Annals of Botany*, **27**, 365–369.

Roberts, E. H. (1964a). The distribution of oxidation-reduction enzymes and the effects of respiratory inhibitors and oxidising agents on dormancy in rice seed. *Physiologia Plantarum*, **17**, 14–29.

Roberts, E. H. (1964b). A survey of the effects of chemical treatments on dormancy in rice seed. *Physiologia Plantarum*, **17**, 30–43.

Roberts, E. H. (1965). Dormancy in rice seed. IV. Varietal responses to storage and germination temperatures. *Journal of Experimental Botany*, **16**, 341–349.

Roberts, E. H. (1972). Cytological, genetical and metabolic changes associated with loss of viability. In *Viability of Seeds* (ed. E. H. Roberts), pp. 253–306. Chapman and Hall, London.

Roberts, E. H. and Benjamin, S. K. (1979). The interaction of light, nitrate and alternating temperature on the germination of *Chenopodium album, Capsella bursa-pastoris* and *Poa annua* before and after chilling. *Seed Science and Technology,* **7**, 379–392.

Roberts, E. H. and Smith, R. D. (1977). Dormancy and the pentose phosphate pathway. In *The Physiology and Biochemistry of Seed Dormancy and Germination* (ed. A. A. Khan), pp. 385–411. Elsevier/North-Holland Biomedical Press, Amsterdam.

Sikder, H. P. (1967). Dormancy of paddy seeds in relation to different seed treatments. *Experimental Agriculture,* **3**, 249–255.

Singh, P. V., Singh, M. B. and Khanna, A. N. (1973). Note on the temperature requirements of germinating rice seeds under controlled conditions. *Indian Journal of Agricultural Science,* **43**, 426–427.

Subramoney, N. and Abraham, A. (1969). A new method of seed treatment against bacterial leaf blight. *International Rice Community News Letter,* **18**, 33–34.

Sugawara, T. (1973). On the dormancy of seeds in *Oryza glaberrima. Bulletin of the College of Agriculture, Utsunomiya University, Japan,* **8**, 43–49.

Tokumasu, S., Kato, M. and Yano, F. (1975). [The dormancy of seed as affected by different humidities during storage in *Brassica*.] *Japanese Journal of Breeding,* **25**, 197–202.

Tseng, S. (1964). Breaking dormancy of rice seed with carbon dioxide. *Proceedings of the International Seed Testing Association,* **29**, 445–450.

Vincent, E. M. and Roberts, E. H. (1977). The interaction of light, nitrate and alternating temperature in promoting the germination of dormant seeds of common weed species. *Seed Science and Technology,* **5**, 659–670.

Vincent, E. M. and Roberts, E. H. (1979). The influence of chilling, light and nitrate on the germination of dormant seeds of common weed species. *Seed Science and Technology,* **7**, 3–14.

Yasue, T. (1973). [Effect of moisture content of seeds and soaking in water on breaking dormancy in *indica* rice.] *Research Bulletin of the Faculty of Agriculture, Gifu University,* **34**, 1–10.

Yasue, T. and Asai, Y. (1970). [The influence of storage conditions on dormancy and germination in rice seed.] *Research Bulletin of the Faculty of Agriculture, Gifu University,* **29**, 67–79.

Mathur, S. B. (1983), *Seed Sci. & Technol.* **11**, 113–128

Testing seeds of tropical species for seed-borne diseases

S. B. MATHUR

Danish Government Institute of Seed Pathology for Developing Countries, Ryvangs Alle 78, DK-2900 Hellerup, Copenhagen, Denmark

(Accepted September 1982)

Summary

Techniques for testing seeds of tropical crops for many seed-borne diseases are now available. The article reviews the pioneer role the Plant Disease Committee of the International Seed Testing Association has played in developing these methods. Contributions made by other institutions, within the tropics and outside, including efforts made at national and international levels in training scientists and technicians, are presented. Routine testing of seed for seed-borne diseases in seed production programmes and plant quarantine inspection services is conducted presently on an extremely limited scale because the testing laboratories lack facilities and trained staff.

Résumé

Analyse des maladies transmises par les semences chez les semences tropicales

Des techniques d'analyse de nombreuses maladies transmises par les semences chez les semences d'espèces tropicales sont maintenant disponibles. L'article passe en revue le rôle de pionnier que le Comité de Maladies des Plantes de l'Association internationale d'Essais de Semences a joué en développant ces méthodes. On présente les contributions apportées par d'autres organisations, dans les régions tropicales et à l'extérieur, y compris les efforts faits au niveau national et international pour former des scientifiques et des techniciens. Les analyses de routine des semences pour les maladies transmises par les semences dans les programmes de production de semences et les services d'inspection de quarantaine sont menées actuellement à très petite échelle parce que les laboratoires d'essais manquent d'installations et de personnel entraîné.

Zusammenfassung

Die Prüfung von Saatgut tropischer Arten auf samenbürtige Krankheiten

Es stehen jetzt Techniken zur Prüfung von Saatgut tropischer Arten für viele samenbürtige Krankheiten zur Verfügung. Diese Arbeit gibt einer Überblick über die Pionierrolle, welche der Ausschuß für Pflanzenkrankheiten der Internationalen Vereinigung für Saatgutprüfung bei der Entwicklung dieser Methoden gespielt hat. Beiträge anderer Institutionen innerhalb und außerhalb der Tropen, welche auch die Bemühungen zur Ausbildung von Wissenschaftlern und Technikern auf nationaler und internationaler Ebene miteinschließen, werden dargestellt. Zur Zeit wird die Routineprüfung auf samenbürtige Krankheiten innerhalb der Saatgutproduktionsprogramme und der Untersuchungsdienste der Pflanzenquarantäne nur in äußerst begrenztem Umfang durchgeführt, weil es den Prüfungslaboratorien an Ausrüstung und ausgebildeten Mitarbeitern fehlt.

Introduction

Objectives and basic requirements of seed testing for seed-borne diseases, including detailed descriptions of available test methods, have been presented in recent publications (Neergaard, 1979; de Tempe and Binnert, 1979). These references are valuable and important for anyone who wishes to organise routine testing either in seed testing laboratories or quarantine inspection services.

Seed testing in the tropics is generally considered to be in its infancy. I intend to trace the developmental work concerning seed health testing in this part of the world which is mostly occupied by the developing countries.

1. Development through the International Seed Testing Association

When one studies the literature on the history of seed testing for seed-borne diseases it becomes apparent that all investigations were initially made and concentrated on crops of temperate climate in the northern hemisphere. The first outstanding publication, 'Manual for the Determination of Seed-borne Diseases', was published in 1938 by Dr. L. C. Doyer, the Mycologist of the Official Seed Testing Station of Wageningen, the Netherlands. The manual deals with the detection methodology, identification and detailed descriptions of some important disease producing organisms and improving the sanitary conditions by cleaning or treating. The main objective of this publication, which is provided with good quality illustrations, was to provide the essential, fundamental information needed to evaluate the sanitary condition of seed lots by plant pathologists and seed analysts working in seed testing stations. This publication was the outcome of a joint cooperative work of the members of the Committee for the Determination of Seed-borne Diseases of the International Seed Testing Association (ISTA). The Committee is at present called the Plant Disease Committee (PDC) which organises international workshops on seed pathology with the sole purpose of establishing testing methods for seed-borne pathogens which are easy to perform in seed testing laboratories, and which at the same time give reproducible results. Any agreement on the choice of testing techniques can be obtained only by comparison of methods and by checking mutually the interpretation of results. Obviously, in all the workshops the first step was to organise comparative tests for specific organisms, basing the test on defined procedures thoroughly described in all details, including the conditions of the recording and, secondly, to check the results obtained independently by the participants, in joint laboratory working sessions, where duplicate samples, set up according to the methods having been used, can be studied and the possible causes of discrepancies discussed. The results of the testing schemes have been collected in reports, each of which is presented during the subsequent workshop.

In an attempt towards establishing uniformity in seed health testing for tropical crops a series of comparative tests on rice (*Oryza sativa* L.) and maize (*Zea mays* L.) was initiated in 1959 under the chairmanship of Dr Paul Neergaard, PDC, ISTA. In a review article on seed pathology, international co-operation and organisation,

Neergaard (1970) reported that rice seed lots had been tested in six testing schemes and maize in three. These testing schemes, in my opinion, were the basis for establishing a sound background for the standardisation of seed health testing methods for tropical plants.

The increasing and expanding activities of the ISTA Plant Disease Committee were reviewed and reorganised at the Thirteenth International Workshop on Seed Pathology held in Copenhagen, 1972. A number of working groups were formed, including the one on 'Tropical and subtropical crops' with responsibilities to standardise seed health testing methods for major crops (Neergaard, 1975). The working group has a membership of 29 from 16 countries and is concentrating firstly on major cereals and their economically important seed-borne diseases (table 1). Some of the pathogens have been tested in several testing schemes while others in few or once only.

Seed health testing methods have been developed empirically. The method which gives maximum infection counts is considered best, provided it gives consistent results, and on which agreement has been reached by the participating stations. Discussions held at the Seventeenth International Workshop on Seed Pathology suggest that the blotter method should be recommended to test rice seed for *Drechslera oryzae* (Van Breda de Haan) Subramanian & Jain, *Pyricularia oryzae* Cav., and *Trichoconiella padwickii* (Ganguly) Jain. We expect to communicate this recommendation to the Executive Committee of ISTA for inclusion in the International Rules for Seed Testing at the next ISTA Congress.

1.2. Publications from ISTA

ISTA Handbook on Seed Health Testing: TESTING METHODS

There have been three publications: Doyer (1938), de Tempe (1961) and de Tempe and Binnerts (1979). When one reviews the papers on seed health testing from the tropical world one often finds these references quoted. This is also expected as many of the crops grown in temperate countries are also cultivated in the hilly tracts and in the plains of the tropics (winter months). Many of the host-pathogen combinations are the same in the two geographical areas. A method may need slight change here and there according to the locally available testing facilities and specific requirements of pathogens in relation to environmental conditions.

ISTA Handbook on Seed Health Testing: WORKING SHEETS

One of the outstanding achievements of the international co-operative work within the framework of the international workshops on seed pathology is the publication of working sheets (tables 2 and 3). A working sheet gives details of methods which can be used for detecting the specific pathogen for which the sheet is prepared, including essential photographs of the pathogen and of symptoms which are necessary for the identification work; relevant references are cited in order to enable the user to seek further information. In some of the working sheets the methods have been evaluated with regard to reproducibility and agronomic value of the testing results. It

Table 1. Work conducted by the working group on tropical and subtropical crops.

Crop	Seed samples	Pathogens	Method	Year of testing
Zea mays	8	*Cephalosporium acremonium* *Drechslera maydis*	Blotter Deep-freeze blotter	1972, 1975
Oryza sativa	4	*Drechslera oryzae* *Pyricularia oryzae* *Trichoconiella padwickii*	Blotter Deep-freeze blotter Guaiacol agar	1973
	4	*Drechslera oryzae* *Trichoconiella padwickii*	Blotter Guaiacol agar	1977
	4	*Pyricularia oryzae*	Blotter Deep-freeze blotter	1978, 1980
Triticum aestivum	5	*Drechslera sorokiniana* *Septoria nodorum*	Deep-freeze blotter PDA-oxgall Modified PDA-oxgall Test tube seedling symptom test	1975
	1	*Septoria nodorum*	Deep-freeze PDA-oxgall	1977
Sorghum bicolor	1	*Colletotrichum graminicola*	Blotter 2,4-D blotter	1976
Eleusine coracana	3	*Drechslera nodulosa* *Pyricularia grisea*	Blotter Deep-freeze	1976
Pennisetum typhoides	4	*Sclerospora graminicola*	Rolling of seed on paper Seed washing test Embryo count	1977
	2	*Sclerospora graminicola*	Seed washing test Tetrazolium test Embryo count	1978
	2	*Sclerospora graminicola*	Seed washing test Embryo count	1981
Glycine max	1	*Phomopsis sojae*	Deep-freeze blotter 2,4-D blotter PDA	1980

Table 2. Working sheets on seed-borne pathogens issued by the International Seed Testing Association, 1966.

Host	Pathogen	Sheet number
Avena sativa	*Fusarium nivale*	1
Avena sativa	*Fusarium avenaceum* and *F. culmorum*	2
Avena sativa	*Drechslera avenae*	3
Brassica spp.	*Alternaria brassicicola*	28
Brassica spp.	*Alternaria brassicae*	29
Brassica spp.	*Sclerotinia sclerotiorum*	30
Brassica spp.	*Phoma lingam*	31
Daucus carota	*Alternaria dauci*	4
Daucus carota	*Stemphylium radicinum*	5
Glycine max	*Diaporthe phaseolorum*	27
Hordeum spp.	*Drechslera graminea*	6
Hordeum spp.	*Drechslera teres*	7
Hordeum spp.	*Drechslera sorokiniana*	8
Hordeum spp.	*Drechslera siccans*	17
Hordeum vulgare	*Ustilago nuda*	25
Lactuca sativa	Lettuce mosaic virus	9
Linum usitatissimum	*Botrytis cinerea*	10
Oryza sativa	*Drechslera oryzae*	11
Oryza sativa	*Pyricularia oryzae*	12
Oryza sativa	*Trichoconiella padwickii*	13
Oryza sativa	*Drechslera halodes*	20
Oryza sativa	*Curvularia lunata* and *C. spicifera*	21
Oryza sativa	*Nigrospora oryzae*	22
Oryza sativa	*Epicoccum* spp.	24
Papaver somniferum	*Dendryphion penicillatum*	14
Pisum sativum	*Ascochyta pisi*	16
Sesame orientale	*Sclerotium bataticola*	26
Triticum spp.	*Drechslera avenacea*	18
Triticum spp.	*Septoria nodorum*	19
Triticum spp.	*Fusarium nivale*	33
Zinnia elegans	*Alternaria zinniae*	32

is hoped that they will be useful for plant pathologists and seed analysts working in routine seed testing as well as for those working in research as almost all the listed organisms have been reported from the tropics. The sheets can be bought from the ISTA Secretariat in Zürich, Switzerland.

ISTA Handbook on Seed Health Testing: AN ANNOTATED LIST OF SEED-BORNE DISEASES

This handbook has been published thrice: the first time in 1959 (Noble, de Tempe and Neergaard: 159 pages), the second in 1968 (Noble and Richardson: 191 pages) and the third in 1979 (Richardson: 320 pages), with a Supplement 1 published in 1981 (78 pages). The hosts are arranged alphabetically irrespective of their botanical taxonomic positions. Under each host, references are listed first on fungi, followed by

Table 3. Working sheets on seed-borne pathogens issued by the International Seed Testing Association, 1981.

Host	Pathogen	Sheet number	Edition
Beta vulgaris	*Phoma betae*	49	1st
Cicer arietinum	*Ascochyta rabiei*	38	1st
Coriandrum sativum	*Protomyces macrosporus*	41	1st
Crucifereae	*Xanthomonas campestris*	50	1st
Eleusine coracana	*Drechslera nodulosa*	36	1st
Eleusine coracana	*Pyricularia grisea*	37	1st
Glycine max	*Cercospora kikuchii*	34	1st
Glycine max	*Diaporthe phaseolorum*	27	2nd
Gossypium spp.	*Colletotrichum gossypii*	42	1st
Gossypium spp.	*Xanthomonas malvacearum*	43	1st
Helianthus annuus	*Botrytis cinerea*	44	1st
Hordeum vulgare	*Ustilago nuda*	25	2nd
Lactuca sativa	Lettuce mosaic virus	9	2nd
Linum usitatissimum	*Alternaria linicola*	46	1st
Linum usitatissimum	*Botrytis cinerea*	10	2nd
Linum usitatissimum	*Phoma exigua*	47	1st
Lolium spp.	*Gloeotinia temulenta*	35	1st
Oryza sativa	*Pyricularia oryzae*	12	2nd
Pennisetum typhoides	*Sclerospora graminicola*	40	1st
Phaseolus vulgaris	*Colletotrichum lindemuthianum*	45	1st
Sesamum indicum	*Macrophomina phaseolina*	26	2nd
Triticum aestivum	*Ustilago nuda*	48	1st
Zea mays	*Drechslera maydis*	39	1st

bacteria, viruses and nematodes covering topics ranging from seed-borne nature, detection, identification, distribution, surveys, losses and control measures. This publication is a great asset to all, including scientists and technicians in the tropical world and must be available in the testing laboratories.

1.3. Training Courses

Participation in the comparative seed health testing schemes, organised by PDC ever since 1958, has been educative in testing crop seed for different infections. This is a general feeling expressed by participants attending the international workshops. So far, 17 such workshops have been held, each having attendance ranging from 15 to 80.

Educational activities of the Plant Disease Committee were strengthened when the First Teaching Course in Seed Health Testing, attended by 11 participants from 10 countries, was organised in 1976 at the Official Seed Testing Station for England and Wales, Cambridge, United Kingdom. The aims of the course were to teach, in detail, routine methods for the detection of fungal pathogens, discussing advantages and disadvantages of particular procedures, to demonstrate points of discrepancy between stations and to discuss the level of uniformity of results obtainable in practice and ways of checking results. With the same objectives the Second Teaching Course, attended by 11 participants from 11 countries, was given at the Seed Testing Station of Lisbon, Portugal in 1979. In both the courses scientists and technicians working on seed-borne diseases of tropical plants participated. The third course, planned for 1982 in Poland, could not be held because of the local situation. These training courses have been appreciated and that is the reason why PDC has decided to hold them regularly in future.

1.4. The Danish Government Institute of Seed Pathology for Developing Countries

Since its birth in 1967 and until today (August 1982), the Danish Institute of Seed Pathology in Copenhagen has trained 259 plant pathologists, seed scientists and technicians from 58 developing countries of the tropical world in the field of seed pathology. The trainees receive training in seed health testing methods for detecting seed-borne diseases caused by bacteria, fungi, nematodes and viruses. The training is imparted mostly using seed samples of different crops from the tropics brought to Denmark by the trainees themselves. The testing is conducted in such an extended manner so that the trainees can start seed testing for seed-borne diseases independently in stations and plant quarantine laboratories on return to their home countries. Emphasis is placed on the philosophies and criteria used in assessing seed-borne diseases for planting value, seed certification, quarantine and seed treatment. The testing methods are also evaluated for their efficiency considering availability of the testing facilities in the developing countries, including technical help available for conducting routine testing work.

Some of the major problems faced in seed health testing of tropical seeds have been the identification of associated organisms, distinction between the harmful and unharmful species and how quickly they can be diagnosed. Research papers published from the Institute on the identification of some of the common genera of fungi, such as *Curvularia* (Benoit and Mathur, 1970), *Fusarium* (Ram Nath, Neergaard and Mathur, 1970), *Drechslera* (Chidambaram, Mathur and Neergaard, 1973), *Myrothecium* (Nguyen, Neergaard and Mathur, 1973), *Colletotrichum* (Kulshrestha, Mathur and Neergaard, 1977) are useful references. An outstanding contribution on the identification and diagnosis of seed-borne plant viruses was made by Phatak (1974).

The Danish Institute of Seed Pathology has organised four regional workshops on seed pathology. The first one was held in the Philippines (1969, attended by 21 participants from 10 countries of the East and Far East), the second in Ghana (1970, attented by 22 participants from 11 countries of Africa, South of the Sahara), Turkey (1972, 17 participants from nine countries of Africa, North of the Sahara) and Argentina (1973, attended by 20 participants from 10 countries of Latin America). These workshops were of two weeks duration and one of the objectives was to introduce the participants to the seed health testing methods. The participants came from different institutions, such as universities, research organisations, seed testing laboratories, seed production agencies and plant quarantine laboratories. A special meeting for the quarantine personnel of Africa and Asia was held in 1980 in India. It was attended by 26 participants from 24 countries. The seminar on university teaching of seed pathology for Africa and Asia, held in India and attended by 26 delegates from 13 countries, included discussion on how to teach various aspects of seed pathology, including seed health testing at undergraduate, graduate, and technician levels.

2. Development in the tropics through local efforts

In tracing the history in the tropics it appears that in the early 1950s, seeds of barley (*Hordeum vulgare* L.), rice, maize, wheat (*Triticum aestivum* L.), sorghum (*Sorghum* spp.), millet, soybean (*Glycine max* (L.) Merr.), peanut (*Arachis hypogaea* L.), and cotton (*Gossypium* sp.) were tested in China for seed-borne diseases in quarantine laboratories (Mrs. Z. Zhàng, personal communication). The testing work was confined mostly to the detection of fungi and in some cases of bacteria and nematodes, using general plant pathological techniques available at that time. However, today, for the international quarantine, seed health testing is conducted for bacteria, fungi, nematodes and viruses using common seed health testing methods, semi-selective and selective media, and sophisticated techniques utilising bacteriophages and serology.

In 1956, a seed health testing unit established at the Mycology Division of the Indian Agricultural Research Institute in New Delhi, India, started testing seeds for their sanitary condition under the leadership of Dr. D. Suryanarayana. A survey made on seed samples of 10 common vegetables revealed the presence of species of *Alternaria, Aspergillus, Colletotrichum, Chaetomium, Curvularia, Fusarium, Gloeosporium, Drechslera (Helminthosporium), Penicillium, Phomopsis, Rhizopus, Rhizoctonia and Stemphyllium* (Suryanarayana and Bhombe, 1961). Identification was done up to generic level only and comments were provided on seed abnormalities, fungi causing them, and on those reducing germination and seedling diseases. Isolation of these fungi was carried out on potato dextrose agar (PDA), malt agar (MA) and moist filter paper, besides examining the seed under binocular microscope (20 $\times$) for discolouration and change in seed morphology. Malt agar was found to be the best medium for isolating many of these fungi. Testing of vegetable seed was also the theme in the second publication from this laboratory (Suryanarayana, Bhombe and Ram Nath, 1963). It is clearly mentioned by these authors that the methods adopted in these investigations were generally those approved by the International Seed Testing Association (Doyer, 1938). In a few cases, slight modifications were made before using them according to locally available laboratory testing conditions.

The laboratory in India has played a key role in studying seed-borne diseases of tropical crops. This is evident from a series of publications that appeared from this laboratory on maize (Dhanraj and Mathur, 1965), fingermillet (*Setaria italica* (L.) Beauv.) (Grewal and Pal, 1965; Mathur, Goel and Joshi, 1967), sorghum (Mathur, Sharma and Joshi, 1967) and peanuts (Lal and Mathur, 1967). It will be impossible to list here the many publications which have appeared from India since then. However, a research bulletin, consisting of 146 pages, on seed-borne fungi and viruses of some important crops (Agarwal, 1981) should act as a good reference source. This bulletin describes testing methods including control measures.

Seed pathology projects on national levels are being planned and organised in some countries. In India, an All India Co-ordinated Project on Seed Pathology, with eight centres under the Indian Council of Agricultural Research, has been established in 1981. An ambitious, co-ordinated research programme on seed pathology was launched in Brazil, in 1978, with 93 projects involving 137 scientists. The main objectives are to carry out research, training, exchange and dissemination of technology in the detection and control of seed-borne diseases (*Seed Pathology News*, No. 11, 1978).

In an effort to train more and more people in seed health testing, the developing countries have been organising short courses and workshops. Ten such courses have been held in India between 1967–1980 and one each in Costa Rica (1976), Brazil (1977), Nigeria (1979) and South Korea (1981). Most of these courses were organised in cooperation with the Institute of Seed Pathology, Copenhagen.

3. Seed health testing practices

In the preceding sections a review has been presented on research and on development of testing techniques for seed-borne diseases of tropical countries, including training of scientists and technicians who should perform the testing in practice. Let us now examine how the testing is conducted in the control laboratories.

3.1 In seed production

As compared to the technically advanced countries, seed production in the tropics is still in its early developmental phases. Seed certification schemes do not exist in most of the countries, while in others only a start has been made on the national level. However, in some countries private seed companies have emerged. Whatever may be the form of the seed certification, testing seed for quality includes purity analysis, germination, in general moisture determination, and trueness to variety. Seeds are seldom tested routinely for seed-borne diseases. The following factual information from a few countries will illustrate this observation.

The Philippines has 11 regional seed testing laboratories and seed certification is presently conducted on some of the major crops. Although all the laboratories are staffed with a trained seed health analyst, only seed lots showing poor germination are tested for seed-borne diseases. In rice, many seed lots, infected with the Bakanae disease organism, *Fusarium moniliforme* Sheld., show low germination. The germination percentages are improved when the seed is treated with Dithan M-45 or Benlate-T. Most of the testing laboratories have formulated research programmes where they would like to establish correlations between seed health test results and the performance of seed lots in the field. This information was gathered during the author's visit to the Philippines in 1981. A simular situation is found in Uganda where a trained seed pathologist is appointed at the Seed Testing Laboratory but only those seed lots are tested for seed-borne diseases which show low germination.

In India, there are 42 seed testing laboratories and according to my information none of them carry out routine seed health tests as they are not manned with trained seed health testing analysts. Some state laboratories take help from plant pathologists working with universities, research institutions and government agricultural departments in evaluating the health status of some seed lots, including advice on seed treatments.

The seed testing laboratory of the National Seeds Corporation (NSC) has a seed health testing staff of five officers and four technicians. In 1977, certified wheat seed lots were tested for loose smut (*Ustilago nuda* (Jens.) Rostr.) infection by the embryo count procedure. A tolerance level of 0.5% was fixed. All seed lots above this level were treated with the systemic fungicide Vitavax to control the smut before the wheat seed was made available to the farmer. As a result, loose smut infection in the field was negligible in 1977 (Dr. A. K. Lambat, New Delhi personal communication). The testing scheme cost 50.000 Indian Rupees but it saved 3 million Rupees in seed treatment alone. Unfortunately the laboratory testing was discontinued in 1978. Field inspection was followed and relied upon and seed treatment considered unnecessary for those seed lots that had passed through this scheme. As a result of this policy the disease incidence increased and infection of loose smut as high as 4% was observed in some fields during 1978. The total loss of the farmer amounted to 0.3 million tons. Seed treatment of certified wheat has now become a part of the certification scheme. A similar project was undertaken at G. B. Pantnagar University of Agriculture and Technology, Pantnagar. Seed health testing saved 2 million Rupees as many seed lots did not need Vitavax treatment against loose smut.

The Seed Certification Laboratories in Nepal do not submit seed lots for routine examination of seed-borne diseases. But, realising the heavy losses caused by loose smut disease in wheat all certified wheat in the country is treated with Vitavax. The fungicide was provided by Canada to Nepal as gift for a few years, but the aid was stopped in 1981. As a consequence seed treatment of certified wheat would now cost US $ 105,000 to the Agricultural Input Corporation of Nepal (Dr. Badri Kayastha, personal communication). This heavy expenditure can be reduced if wheat seed lots are examined routinely for loose smut infection and seed treatment given only to those carrying infection higher than the tolerance level permitted.

Seed certification for wheat, barley and oats (*Avena sativa* L.) based on field inspection was introduced in Morocco as far back as 1960. But it seems that smuts are still causing heavy losses. Lyamani (1975) tested certified and non-certified seed lots by laboratory seed health testing procedures and demonstrated that even the certified lots were infected considerably by smuts and bunts, indicating the need for laboratory seed testing by proper techniques. This investigation was conducted as a part of a research programme, but we do not know whether the seed lots are now routinely tested for seed-borne diseases in the country. Similar situations are found in other countries.

In Bangladesh, jute (*Corchorus* spp.) is the only crop on which any appreciable amount of testing has been done. The basic seed is tested for jute mosaic and important fungal diseases at the Central Jute Research Institute, Dacca.

3.2. In plant quarantine

Due to risks involved in introducing dangerous plant diseases through seed to new areas, testing seed for seed transmitted diseases has received greater attention by the plant quarantine organisations. As a result, routine examination of commercial seed lots and germplasm has been introduced in some countries of Africa, Asia and Latin America. Most of the testing is, however, confined to the detection of seed-borne nematodes and fungal pathogens, particularly by following examination of dry seed, seed washings and growing-on tests for fungi in the past and, more recently, additionally by incubation methods. Testing for bacteria and viruses has been practised only on a very limited scale. Examples given below will depict the actual situation in some of the countries.

The quarantine laboratories in Africa, for example the one at Muguga in Kenya and the other at Ibadan, Nigeria where thousands of seed samples are tested each year, deserve special mention. These laboratories are the leading institutions having trained seed pathologists and seed health analysts. The Inter African Phytosanitary Council (IAPC) of the Organization of African Unity (OAU) is at present planning to establish six regional plant quarantine stations for the whole continent.

The National Bureau of Plant Genetic Resources (NBPGR) situated in New Delhi tests 50,000–60,000 seed samples annually while the Central Plant Protection Training Institute in Hyderabad examines 35,000 samples each year for export and import. In Thailand, a quarantine seed health testing laboratory has been in operation for a number of years.

In Latin America, testing of seed for dangerous diseases has been undertaken in Brazil, Colombia and Mexico. The work load of the national quarantine services of these countries has increased substantially in the last few years, primarily because of the exchange of genetic material. Naming of a few countries in this article does not imply that testing is not done in other countries.

4. Constraints

4.1. Trained staff and testing facilities

During the FAO/DANIDA Seminar on Quarantine for Seed held in New Delhi, India, 1980, all the delegates from the 23 participating countries indicated that they could not test seed for different types of infections effectively and efficiently as their laboratories were not staffed with trained personnel and the testing facilities were either inadequate or missing altogether (table 4). Since 1967, I have visited many seed testing stations in different countries of the tropics and according to my observation the situa-

Table 4. Information collected from country representatives during the FAO/DANIDA Seminar on Quarantine for Seed, held in New Delhi, 1980.

Country	Facilities available	Facilities required
Bangladesh	7 plant quarantine check posts, (2 seaport, 2 airport, 3 landfrontiers).	Seed health testing laboratory, post-entry quarantine facilities, training of staff.
China	36 plant quarantine stations, sufficient equipment and other facilities, few seed health testing laboratories.	More seed health testing laboratories, post-entry quarantine facilities, training of staff.
Egypt	Well equipped laboratory, trained technicians	Seed health testing laboratory and other facilities required at plant quarantine station at Alexandria.
Ethiopia	3 plant quarantine stations.	Seed health testing laboratory, training of staff members.
Fiji	4 plant quarantine stations, fumigation chamber, inspection facilities, post-entry quarantine station will be available.	Seed health testing laboratory, training of staff members.
Ghana	1 seed health testing laboratory.	Fumigation equipment, post-entry quarantine station and training of staff.
India	24 plant quarantine stations, total staff 300, 35 officers including 16 plant pathologists, only three officers trained as seed pathologists, work carried out in collaboration with National Bureau of Plant Genetic Resources.	Additional facilities, training of staff members.
Indonesia	35 plant quarantine stations, 2 seed health testing laboratories, 5 screen houses.	More post-entry quarantine stations, seed health testing laboratories, training of staff members.
Kenya	Inspection facilities, post-entry quarantine facilities, tissue culture laboratory.	Seed health testing laboratory, training of staff members.
Korea	5 branch officers, 110 persons, two trained seed pathologists, 6 seed health testing laboratories.	Two post-entry quarantine stations, laboratory facilities, training of staff members.
Libya	Inspection and treatment available at most ports of entry.	Lack of trained personnel in seed health testing, post-entry quarantine facilities.
Malawi	Seed health testing laboratory, post-entry quarantine station, entry check posts.	Virus testing facilities, more trained seed pathologists.
Malaysia	Insect proof house, laboratories at 3 plant quarantine stations.	Equipment and training of staff, post-entry quarantine stations.
Mauritius	Post-entry quarantine greenhouse, fumigation facilities.	Seed health testing facilities, urgent training in special disciplines of seed health testing.
Nepal	Fumigation facilities, seed health testing laboratory, 7 plant quarantine check posts.	Strengthening of plant quarantine and seed health testing laboratory.

Table 4. (Continued).

Country	Facilities available	Facilities required
Nigeria	2 seed health testing laboratories, 1 for import, 1 for export.	More training required in such specialised areas as virology and bacteriology.
Pakistan	Seed health testing laboratory, 2 trained persons.	Post-entry quarantine facilities, training of staff members.
Philippines	Inspection and fumigation facilities available at post-entry in Manila only.	Seed health testing laboratory attached to plant quarantine office and post-entry quarantine, training of staff members.
Somalia	No facilities available.	Well equipped laboratories, also for seed health testing, trained staff.
Sri Lanka	Laboratories and greenhouse facilities, post-entry quarantine facilities, 2 greenhouses.	Training of staff for seed health testing.
Thailand	1 seed health testing laboratory, 23 plant quarantine stations, 1 post-entry quarantine station.	Trained personnel, two more post-entry quarantine stations, including seed health testing laboratories.
Uganda	Phytosanitary inspection at international airport, main post office and at border stations (but no facilities).	Laboratories and trained staff.
Zambia	1 plant quarantine station will be opened soon, 1 check post.	Seed health testing laboratory, training of staff members.

tion in these stations is just the same as pointed out in table 4 for quarantine organisations. One of the major problems in the developing countries is the non-availability of funds for establishing proper equipment for testing. Also, publications dealing with details on the physical set-up of seed health testing laboratories for testing bacteria, fungi, nematodes and viruses are lacking. The Institute of Seed Pathology has often been contacted on this issue and it is our intention to prepare a pictorial guide which can serve this purpose.

It must be emphasised that the testing laboratories must utilise local resources and the testing techniques be adjusted accordingly. Two examples may be presented here. Use of near ultraviolet light (NUV), which promotes sporulation of many seed-borne fungi in the blotter method, has become a standard procedure in most of the laboratorities in the technically advanced countries. But this light source is difficult to procure in the developing countries and can safely be replaced by white fluorescent light tubes which are freely found in these countries (Mathur and Neergaard, 1972; Agarwal, Mathur and Neergaard, 1972). In the embryo count method for detecting loose smut in barley, a stream of hot water is used to agitate the grains soaked earlier in 5% solution of sodium hydroxide to obtain better release of embryos from the endosperm

(Rennie and Seaton, 1975). A supply of running hot water is practically never available in tropical laboratories. The procedure may be performed satisfactorily in these countries during the summer months when tap water is generally around 25 °–28 °C. The temperature may even be slightly higher in those places where the water is supplied from overhead tanks exposed to sunlight.

The percentage recovery of embryos in this case may be slightly lower but, to compensate this, one can always start with more seed.

Testing of bacteria and viruses involves simple as well as complicated procedures involving expensive chemicals and sophisticated apparatus and instrumentation. Most of the countries will not be able to afford separate laboratory installations of their own. It may be most practical if a well equipped laboratory for detecting these groups of organisms is provided in a central seed health testing laboratory for the whole country or one for each region depending upon the availability of funds.

During the Second Meeting of the Expert Consultation on Seed Industry Development Programme (SIDP), it was concluded that by 1980 there will be a need for 8,500 persons having academic level and 35,000–40,000 people at intermediate level in cereal seed technology alone in the developing countries (Wold, 1975). This estimation was made when 54 countries were participating in the programme. The figures will be extremely high when we consider that 92 countries are now participating. The figures will get further multiplied when other crops are considered. No such calculations have been made so far on the need for seed health testing personnel. Trained personnel will be needed at the academic and technician levels. Scientists having an academic background are important for taking decisions on the planting value in seed certification programmes and for rejecting and accepting seed consignments in plant quarantine services. Generally, the staff appointed at plant quarantine organisations is trained in plant pathology but not those at seed testing stations and therefore more extended and intensive training will be needed in seed health testing procedures.

4.2. Planning

Hundreds of seed production projects at national and international levels are in operation and at the same time many more are being planned today for the tropical countries. In almost all of them laboratory testing of seed for seed-borne diseases has been neglected. It is vital that seed is tested at all stages of multiplication for important seed transmitted diseases. Testing of pre-basic seed (breeders seed) is most important as it is at this stage that the seed, if found infected, can be given effective treatment. Treating small quantities is much easier and cheaper. In fact, diseases which are exclusively seed-borne can be eliminated.

In a number of countries the seed production scientists try to take help from the plant pathologists of agricultural departments, research or teaching institutions; but they are often not able to assist as they are themselves not familiar with routine techniques for detecting different kinds of seed infections, and sometimes they cannot spare time as they are busy with their own daily work. I have witnessed such situations in my travels. Therefore a better co-ordination between outside plant pathologists and seed production staff and, of course, ideally, appointment of trained seed pathologists in seed projects will ensure further production of high quality seed.

Another problem that deserves consideration by the planners is the heavy load of testing work scientists and technicians face today in plant quarantine laboratories, created by the exchange of enormous quantities of genetic material for breeding purposes. Exchange of germplasm is essential for breeding high yielding cultivars but the number of samples passing through a testing laboratory must not exceed what can be tested properly by the existing staff.

References

Agarwal, V. K., Mathur, S. B. and Neergaard, P. (1972). Some aspects of seed health testing with respect to seed-borne fungi of rice, wheat, blackgram, greengram, and soybean grown in India. *Indian Phytopathology*, **25**, 91–100.

Agarwal, V. K. (1981). Seed-borne fungi and viruses of some important crops. *University of Agriculture and Technology, Pantnagar, India Research Bulletin* 108.

Benoit, M. A. and Mathur, S. B. (1970). Identification of species of *Curvularia* on rice seed. *Proceedings of the International Seed Testing Association*, **35**, 99–119.

Chidambaram, P., Mathur, S. B. and Neergaard, P. (1973). Identification of seed-borne *Drechslera* species. *Friesia*, **10**, 165–207.

Co-ordinated Research Programme on Seed Pathology in Brazil. (1978). *Seed Pathology News*, **11**, 9.

de Tempe, J. (1961). Routine methods for determining the health condition of seeds in the seed testing stations. *Proceedings of the International Seed Testing Association*, **26**, 27–60.

de Tempe, J. and Binnerts, J. (1979). Introduction to methods of seed health testing. *Seed Science and Technology*, **7**, 601–636.

Dhanraj, K. S. and Mathur, S. B. (1965). Ear rot of maize caused by *Cephalosporium acremonium* Corda, a new record for India. *Indian Phytopathology*, **18**, 393–394.

Doyer, L. C. (1938). *Manual for the determination of seed-borne diseases*. International Seed Testing Association, Wageningen.

Grewal, J. S. and Pal, Mahendra (1965). Seed microflora II. Seed-borne fungi of *Setaria italica*, their distribution and control. *Indian Phytopathology*, **18**, 123–127.

Kulshrestha, D. D., Mathur, S. B. and Neergaard, P. (1976). Identification of seed-borne species of *Colletotrichum*. *Friesia*, **11**, 116–125.

Lal, Sashi Prabha and Mathur, S. B. (1967). Studies on seed-borne fungi of groundnut. I. A new method for detecting fungi in small samples of seed, with special reference to *Macrophomina phaseoli* (Maubl.) Ashby (Syn. *Sclerotium bataticola* Taub.). II. Laboratory assay of fungicides for controlling seed-borne infections. *Proceedings of the International Seed Testing Association*, **32**, 655–665.

Lyamani, A. (1975). *Etude de la composition de la flora fongique associee aux semences de ble et d'orge au Maroc*. Institut Agronomique et Veterinaire – Hassan II, Rabat.

Mathur, S. B., Goel, L. B. and Joshi, L. M. (1967). Fungi associated with seeds of *Setaria italica* and their control. *Proceedings of the International Seed Testing Association*, **32**, 633–638.

Mathur, S. B., Sharma, R. and Joshi, L. M. (1967). Mouldy head disease of Sorghum; isolation of associated fungi on blotter and agar. *Proceedings of the International Seed Testing Association*, **32**, 639–645.

Mathur, S. B. and Neergaard, P. (1972). Seed health testing of rice. IV. Effect of light and temperature on seed-borne fungi in the blotter test. *Proceedings of the International Seed Testing Association*, **37**, 723–730.

Neergaard, P. (1970). Seed Pathology, international co-operation and organisation. *Seed Science and Technology*, **35**, 19–42.

Neergaard, P. (1975). Some salient points in co-operation within seed pathology. *Advances in Mycology and Plant Pathology,* New Delhi, 241–248.

Neergaard, P. (1979). *Seed Pathology.* Vols 1 and 2. The Macmillan Press Ltd., London.

Nguyen, Thu Ha., Mathur, S. B. and Neergaard, P. (1973). Seed-borne species of *Myrothecium* and their pathogenic potential. *Transactions of the British Mycological Society,* **61,** 347–354.

Noble, M., de Tempe, J. and Neergaard, P. (1959). *An annotated list of seed-borne diseases.* ISTA, Wageningen.

Noble, M. and Richardson, M. J. (1968). An annotated list of seed-borne diseases. *Proceedings of the International Seed Testing Association,* **33**, 1–191.

Phatak, H. C. (1974). Seed-borne plant viruses – Identification and diagnosis in seed health testing. *Seed Science and Technology,* **2**, 3–155.

Ram Nath, Neergaard, P. and Mathur, S. B. (1970). Identification of *Fusarium* species on seeds as they occur in blotter test. *Proceedings of the International Seed Testing Association,* **35**, 121–144.

Richardson, M. J. (1979). *An annotated list of seed-borne diseases.* ISTA, Zürich.

Richardson. M. J. (1981). *Supplement I to An annotated list of seed-borne diseases.* ISTA, Zürich.

Rennie, W. J. and Seaton, R. D. (1975). Loose smut of barley. The embryo as a means of assessing loose smut infection in seed stocks. *Seed Science and Technology,* **3**, 697–709.

Suryanarayana, D. and Bhombe, B. B. (1961). Studies on the fungal flora of some vegetable seeds. *Indian Phytopathology,* **14,** 30–41.

Suryanarayana, D., Bhombe, B. B. and Ram Nath (1963). Testing of vegetable seeds for freedom from diseases. *Indian Journal of Horticulture,* **20,** 141–145.

Wold, A. (1975). Expert consultation on seed industry development, Second Meeting. *ISTA News Bulletin,* **49,** 13–18.

Gregg, B. R. (1983), *Seed Sci. & Technol.*, **11**, 129–148

Seed marketing in the tropics

B. R. GREGG*

Mississippi State University, Seed Technology Laboratory, Mississippi, USA

(Accepted July 1982)

Summary

Marketing seed to small farmers in tropical areas is a highly specialised educational servicing operation, subject to many constrains not the least of which is farmers austere value system. A reliable seed industry base is necessary, with market-oriented organisation and management, supported by an adequate marketing policy. Promotion must be extension educational in nature, complete, honest, intensive and comprehensive. Seed distribution methods and channels must protect seed and get it to farmers when it is needed. Retail seed suppliers are the key individual contact with farmers which convince them to buy and use improved seed. Market intelligence should be used to determine realistic production goals, so that all seed produced can be delivered to farmers. Marketing and promotion requirements, methods, agencies and channels are diagrammed.

Résumé

Commercialisation des semences dans les pays tropicaux

Commercialiser des semences auprès de petits agriculteurs dans les régions tropicales est une opération de service à caractère éducatif hautement spécialisée soumise à de nombreuses contraintes dont le système d'austérité financière des agriculteurs n'est pas la moindre. Une base d'industrie des semences fiable est nécessaire avec une organisation et une gestion de marché soutenues par une politique de commercialisation adéquate. La promotion doit être une vulgarisation de nature éducative, complète, honnête, intensive et cohérente. Les méthodes et les canaux de distribution des semences doivent protéger les semences et permettre leur livraison aux agriculteurs au moment nécessaire. Les détaillants sont la clé des contacts individuels avec les agriculteurs qu'ils convainquent d'acheter et d'utiliser des semences améliorées. La connaissance du marché doit être utilisée pour déterminer des objectifs de production réalistes afin que toutes les semences produites puissent être livrées aux agriculteurs. Les exigences, les méthodes, les agents et les canaux de commercialisation et de promotion sont représentés sous forme de diagrammes.

Zusammenfassung

Saatgutkauf in den Tropen

Der Saatgutkauf ist für kleine Farmer in den Tropen zufolge vieler Zwänge, von denen das strenge Wertsystem des Farmers nicht der geringste ist, ein hochspezialisierter Versorgungsvorgang. Es ist eine zuverlässige Basis der Saatgutindustrie mit einer marktorientierten Organisation und Führung erforderlich, welche

* Present address: Seed Division, Department of Agricultural Extension, Ministry of Agriculture and Cooperatives, 2143/1 Paholyothin Road, Bangkhen, Bangkok, Thailand

durch eine angemessene Marktpolitik unterstützt wird. Die Werbung dafür muß im Wesen weitgehend erzieherisch, vollständig, aufrichtig, intensiv und umfassed sein. Die Saatgutverteilungsmethoden und -kanäle müssen das Saatgut schützen und es zu den Farmern gelangen lassen, sobald es benötigt wird. Der Kleinhandel bildet den Schlüssel für persönliche Kontakte mit den Farmern, um sie zu überzeugen, besseres Saatgut zu kaufen und zu verwenden. Man muß Marktverstand sprechen lassen, um die realistischen Produktionsziele zu bestimmen, damit alles produzierte Saatgut an die Farmer geliefert werden kann. Die Verkaufs- und Werbungserfordernisse, Methoden, Verkaufsstellen und -kanäle werden in Diagrammen dargestellt.

Seed marketing

Highly specialised marketing is required to take improved seed to farmers and get them to use it. As seed production increases, more marketing effort is required to sell the greater quantities of seed.

Marketing brings sellers and buyers together. It brings some order to matching farmers' needs with what is produced, and vice versa. In 'the market' prices are set and output is 'rationed' among buyers. Essential to marketing are:

1. Farmers who want or need to buy the seed: a demand.
2. Someone through whom to sell the seed: a marketing system.
3. Confidence of both seedsman and farmer in the marketing system: value and price, quality, reliability etc.

Marketing includes merchandising, the planning and operations needed to get the right seed to the right farmers at the right time, at the right price, in the right package; selling and promotion, to create farmers' desire and willingness to purchase the seed; distribution, the storage, transport, dealer stocking, etc., required to move seed from producer to the farmer.

Seed marketing ultimately seeks to increase the use of high-yielding crop cultivars through general farmer use of improved seed; seed producers and dealers must gain a profit to encourage them to continue producing seed, but ultimate goals are better income and standard of living for farmers, and increased total national agricultural production.

A major aspect of seed marketing must be to stimulate farmers to produce more, to meet national targets of increased production. It must motivate farmers to produce more to receive income above their basic minimal needs. This, in turn, requires stimulating their desire for more and better goods, services, homes, health care, education, clothing, tools, etc.

In many tropical rural areas, this is not easy; farmers have a value system which constrains use of improved seed and other inputs to increase yields, and ultimately constrains increase of total national production. Traditionally, small tropical farmers lead an austere, simple way of life, with minimum material needs. A minimum increase in income satisfies their frugal material wants; beyond that, they have little incentive to increase their investment and risk so as to produce more to feed the urban population and meet government production targets.

This necessitates broad social changes and communications to create farmer awareness of a better way of life, and stimulate their desire to increase production. Improving their standard of living can happen only by increasing their production and income above minimum needs. Furthermore, the rural society must be secure and stable enough to make the farmer feel safe in spending extra money and owning more goods, to make him willing to invest in improved inputs to produce more.

Some farmers have this personal motivation to increase production and income; however, most small tropical farmers must be stimulated through a combination of government action, social development, education, secure law and order, mass communication and marketing. Seed marketing must be aimed at stimulating and creating among farm families the desire to use improved seed to earn more so they can buy a better standard of living.

Characteristics of tropical seed marketing

Seed marketing has problems and constraints in common with other industries which market to tropical small farmers: how to reach many farmers in many villages, and make them want to buy and use higher-yielding seed.

However, seed is unique in many aspects.

Most rural markets cannot effectively handle seed because of the large volume required, sensitivity of living seed to humidity and temperature, vagaries of production, and the seasonal need for specific seed at specific times in specific areas.

High-yielding seed carries genetic potential for the farmer's crops to produce more, but the appearance of seed does not indicate its quality. The farmer has no easy way of knowing if the seed he buys is really that much better. He must place a great deal of faith in the seed supplier, because its true quality will not be known until it is too late to change. Once a small farmer loses faith in his seed supplier, he will not patronise him again.

Seed marketing must be oriented to farmers' specific personally perceived needs. The seed salesman must know the characteristics of new and improved crop cultivars, be able to give appropriate and complete advice on fertiliser, weed and pest control, crop management and harvest, in a way farmers can relate to and use. He must describe seed quality and its value in relation to crop yields, and give other factual information to create farmer desire to buy the seed.

Time is critical in seed marketing. Seeds are living and will die; they cannot be purchased too far in advance. Because the need for seed exists only at planting time, it must be available at the right time, at the right place, in the right amount, at the right price and of the highest economic quality.

Seed production involves an inflexible time lag of two to four years. Initial decisions are critical, since there is little the seedsman can do to increase volume after the programme has started. Also unique is the number of technical tasks which must be performed at specific critical times, regardless of weather.

Seed marketing must be ethical. The seed producer, processor and dealer must make a reasonable profit. Yet, the farmer must be able to buy high-quality seed at a price

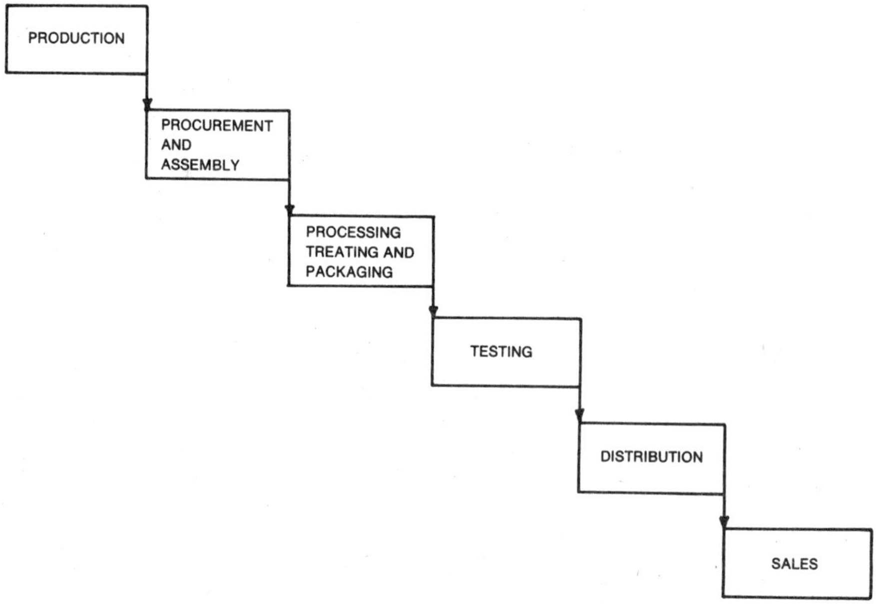

Figure 1. Sequence of physical operations in the seed industry to make improved seed available to farmers.

he can afford, in terms of the profit it will make for him through improved crop yields and quality. Any excess or failure at any point in the system will disrupt the orderly flow of seed from producer to farmer, and reduce the national production potential offered by improved seed. The entire seed industry and its marketing system must be founded on honesty, trustworthiness and reliability.

The seed industry base

Before marketing can be employed to increase farmer use of improved seed, the seed industry must be able to supply enough improved seed.

Adequate amounts of improved seed must be produced and harvested to achieve high quality; properly dried; processed to the desired quality; treated, packaged and held in protective storage until needed for planting. Safe, rapid transport must carry it to points of delivery to farmers, at the proper time for planting. Figure 1 shows the physical chain of operations required to support seed marketing.

Identifying improved seed

The seed programme and its seed must, in the farmer's mind, be unequivocally associated with high quality, trustworthiness, and dependability. Positive, easy identification of improved seed is necessary to promote small-farmer acceptance and to prevent adulteration and misrepresentation. Identification must establish the kind and quality of seed, and its source. Bags should be:

(1) Labelled to prevent mistaken identity;
(2) Marked with a trademark or symbol; and
(3) Sealed to prevent adulteration. Both tags and a stencilled lot number should be on the bag.

Farmers should always receive seed in sealed, unopened bags; bags should not be opened to sell smaller quantities.

Market-oriented management

A seed programme cannot be successful unless its seed is sold to farmers. To ensure an economic return, most successful programmes ultimately find market-oriented operations necessary. Sometimes called the Marketing Concept, this is based on two principles:

1. All planning, policies, production and operations are pointed toward the real needs of farmers: identifying, supplying, servicing and statisfying them.
2. Producing efficiently, promoting effectively, and selling at prices acceptable to farmers, at a volume large enough to permit the necessary return to the programme.

The Marketing Concept is critical to seed industry success. It is really a philosophy of doing business which says that farmers' needs are the justification of the seed programme's existence, so all activities; breeding, research, production, processing, storage, distribution, promotion, marketing, must be devoted to:

1. Determining the wants and needs of farmers; and,
2. Satisfying these while still making a reasonable profit.

Untrained persons have felt that the Marketing Concept applies only to a highly-developed society, not to a developing tropical agriculture. However, the opposite is actually true: the Marketing Concept is more important in a developing agriculture than in a highly-developed situation, because:

1. In a highly-developed agriculture, farmers are educated and already motivated to buy improved seed and produce more. Also, highly-developed marketing channels and agencies do much of the promotional work, so the seedsman can concentrate more on production and distribution.
2. In a developing small-farmer tropical agriculture, farmers must be motivated to buy and use improved seed. They will not go to great effort to seek improved seed even after they have once used it, and they do not understand the principles of improved seed or the requirements for producing it. Marketing channels and outlets are not well-developed; the seed programme must handle its own marketing, train and support its dealers, and carefully produce according to what can be sold.

Failure to apply the Marketing Concept is often seen in developing seed industries: unsold surplus seedstocks, while many farmers still use low-yielding unimproved seed.

To market seed successfully, the seed programme must be totally oriented toward creating farmer demand and then supplying that demand. This starts with programme

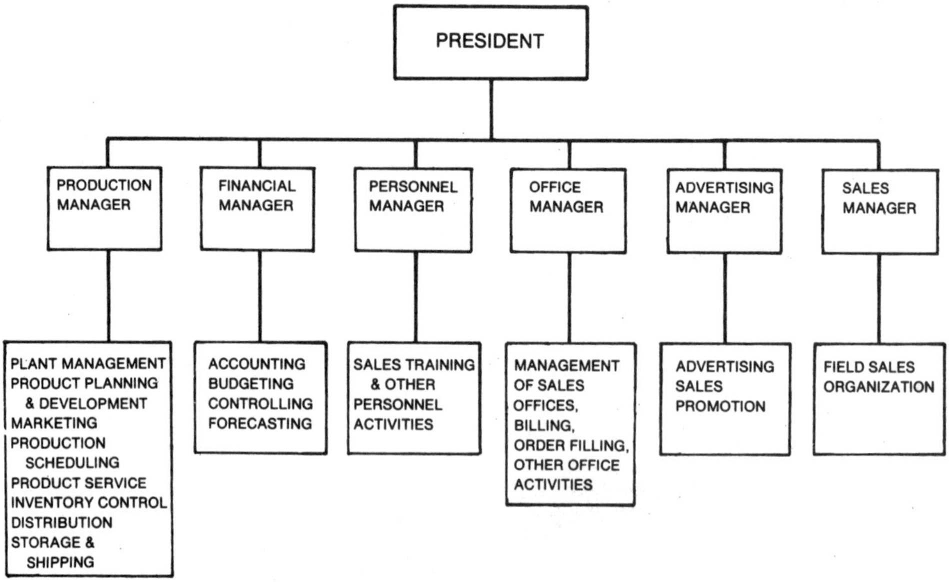

Figure 2. Production-oriented seed programme organisation, which cannot effectively coordinate marketing activities (after Law, Gregg, Young and Chetty, 1971).

organisation, and extends through all planning, management and operations.

In a production-oriented organisation (figure 2) the activities necessary to service the market demand are fragmented, and effective coordination and integration is difficult or impossible. Production, processing and storage cannot be oriented toward the ultimate purpose of selling seed to farmers. Marketing functions placed in different sections make coordination and communication difficult. Diffused responsibilities cause failure to meet farmers' needs, with excesses of some seed and shortages of others. Crises occur constantly, and it is difficult to operate efficiently and market profitably.

In an efficient market-oriented organisation (figure 3), all marketing operations are concentrated under one manager, and can be coordinated to ensure maximum efficiency in creating and meeting farmers' demand. Forecasting farmer demand, planning production, dates and locations of seedstock needs (inventory control of cultivars and amounts), sales promotion, market and sales expansion, dealer development and servicing and other functions can be integrated effectively. Production and processing can then work toward supplying definite, well-planned and realistic market needs. Fewer management crises occur and operations run smoothly. This organisation strengthens production and processing by giving them realistic targets, allowing them to concentrate on technical operations and providing better coordination at the operational management level.

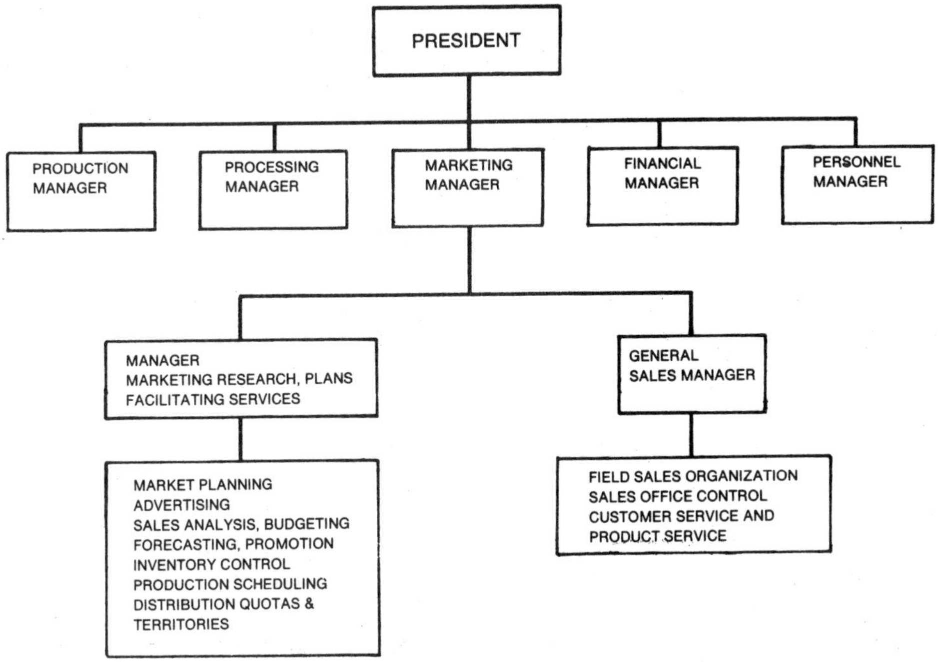

Figure 3. Effective, market-oriented seed programme organisation which can focus all activities toward identifying, developing, and satisfying farmers' demand for improved seed (after Law *et al.*, 1971).

Seed marketing concepts must be integrated into all seed operations. Breeders must ensure better performance of new cultivars and provide data useful in selling seed. Agronomists must provide production data and advice which help sell seed; producers and processors must turn out seed of consistently high quality which maintains farmer trust. Every seed operation and worker must be market-oriented; marketing only presents their results to farmers, in a way which stimulates farmer desire to use improved seed.

Management must be oriented toward the market, supplying the kind and amount of seed that farmers will buy. This requires:

1. Determining goals and objectives: realistic market-oriented long-term goals, with specific short-range goals to implement progress toward long-range goals.
2. Planning operating strategy: both long- and short-range planning must detail the strategy required to meet goals.
3. Organisation and coordination: arrange all activities, facilities and personnel, and designate responsibilities, so as to achieve maximum output and efficiency. Close communication must effectively integrate and coordinate all activities and persons.
4. Staffing, facilities, and funding: sufficient to meet goals and used efficiently.

5. Effective operation and supervision: so that all operations are goal-oriented, efficient, timely, correct and complete.
6. Analysing and evaluating: review all activities critically at regular intervals and improve them as possible.

Marketing policy

The seed programme must have a formal marketing policy which defines objectives and courses of action under given circumstances. The marketing policy should include sales methods, dealer contracts, seed pricing and discount policy, sharing promotion costs with dealers, payment methods, transport, etc. It should establish a policy adequate to guide actions, but not in such detail as to be inflexible. Any change in policy, such as price increases, should be made in off-seasons, when they will not create confusion.

Promotion

Educational level and knowledge of improved seed and crop production is often low among small tropical farmers. Therefore, seed promotional efforts must be intensive and continuous, following the educational extension promotion concept of educating, demonstrating and convincing farmers. All promotion efforts, both within a programme and among all concerned programmes, must be well-coordinated, honest and designed to:

First, make farmers *A*WARE of the existence of improved seed.

Second, arouse farmer *I*NTEREST in what improved seed is and why it should be used.

Third, *D*EMONSTRATE to farmers what improved seed can do under farming conditions they can relate to.

Fourth, *M*OTIVATE farmers to use improved seed for their own crops.

Fifth, *E*DUCATE farmers on how to get improved seed and the credit to purchase it, handle and plant it, and improve their care of the crop it produces.

The AID ME principle must guide all educational promotion to tropical farmers, from the 'innovators' quick to adopt demonstrably improved ideas, to the lower-income groups who are the last to adopt improvements. The majority of small tropical farmers can be convinced to accept and use improved seed if they consistently and effectively learn of its usefulness from sources they trust, see that it can improve income under their own conditions, are helped to obtain it, are constantly kept aware of its value to them, and the seed performs as promised.

Promotion must be comprehensive enough to reach all farmers in the target area, intensive enough to have a positive impact on their choice of seed, and constant and consistent enough to keep them motivated enough to use improved seed every planting season every year. Both government agencies and private sector firms must work together in a carefully planned, coordinated and implemented promotional programme.

Any promotion by any agency should complement that done by all other agencies; it should never conflict or compete with other reliable agencies: the true 'competition' is low-quality seed rather than other brands of improved seed.

Promotion must be educational in nature, to train the farmer to understand, accept and use improved seed; it should not be simply a 'hard sell'.

Promotion must be honest, reliable, and hold the farmer's trust: in the information he receives, the improved seed, its sources, and its value to him personally in his own cropping situation. If he ever has reason to distrust it, or does not receive the promised quality and benefits, he will no longer accept it.

Promotion must be complete, giving the farmer reliable advice not only on the use of improved seed, but also on crops, cultivars, pesticides, irrigation, crop management, harvest, marketing and other concerns with his crops. This not only improves his crop yields and income, but also develops his trust and willingness to accept the guidance of his seed supplier.

Promotion must be intensive, and reach the farmer through every possible source, from mass communications media to personal contacts, every time he has reason to think of his crops. Repetition is as essential to educational marketing to small tropical farmers as it is to any learning process.

Mass media promotion (newspapers, magazines, radio, TV, newsletters, billboards, posters, etc.) makes farmers aware of improved seed, what it is, how it can increase yields and where it is available.

Educational promotion (demonstration farms and plots, cultivar and crop trials, tours and visits to seed operations, field days, training programmes, farmer meetings, club and school projects and contests, youth meetings, etc.), builds farmers confidence in improved seed, shows what it can do under their own crop conditions, and creates the desire to use improved seed. It also directly convinces many farmers to buy and use improved seed.

Individual contacts; by extension specialists and agents, government specialists from other concerned programmes, researchers, seed dealers, merchants, commodity dealers and buyers, etc., convince individual farmers to buy improved seed. The 'point of sale' contact between the seed retailer and farmer is the final and most important individual motivating contact.

The goal of seed promotion, by whatever agency, is to expand farmer use of improved seed and thereby make possible increased production of seed which has an assured market. Promotional efforts must constantly seek to expand the number of farmers who use improved seed. This requires intensive 'market penetration' in the target areas, to get all farmers to use improved seed; it also requires expanding 'geographic coverage' to spread the use of improved seed into larger areas. Intensive promotion can at first be targeted on a relatively small area to reach all its farmers. In later seasons, the acceptance of improved seed established in this area can be maintained with a lower promotional effort, while the intensive promotion programme shifts to another area.

Create and stimulate among farm families:

Desire to produce more, so they can purchase a better standard of living;
Sense of security in buying a better standard of living.

↓

Demonstrate, educate, and promote the value of improved seed as an efficient, sure, economical way to produce more.

↓

Produce adequate quantities of truly improved seed of high genetic yield potential, viability and purity.

↓

Do not wait for farmers to come to you to buy seed; use every possible means of distributing and carrying seed to farmers, wherever they are. Improved seed must be as easily obtainable as alternative kinds of seed.

↓

Make the improved seed easily identifiable.

↓

The improved seed must be dependable and reliable. Farmer faith must be maintained at all times. Even if it means loss at the moment, never permit sale of below-standard seed.

↓

Orient seed operations to farmers' needs. Maintain information feedback not only to produce and supply farmers' real seed needs, but to keep farmers aware that the seed programme is serving them.

Results:
Higher farmer yields
Higher total national production.

Figure 4. The motivational sequence of creating and satisfying demand for improved seed by small tropical farmers.

The educational promotion programme must be complete, well-coordinated and continuous, with regular and adequate budgets and personnel. It must use all suitable promotional, educational and sales methods to publicise the value of improved seed, create farmer desire to use it, and follow through to finalise sales to individual farmers. It must be acceptable to the traditions, customs and social mores of the target farmers, and fit into their economic conditions, while helping them to improve and change for the better. Figure 4 shows the motivational sequence of seed marketing; figure 5 shows the private sector and government agencies which should promote the use of improved seed, and promotional methods.

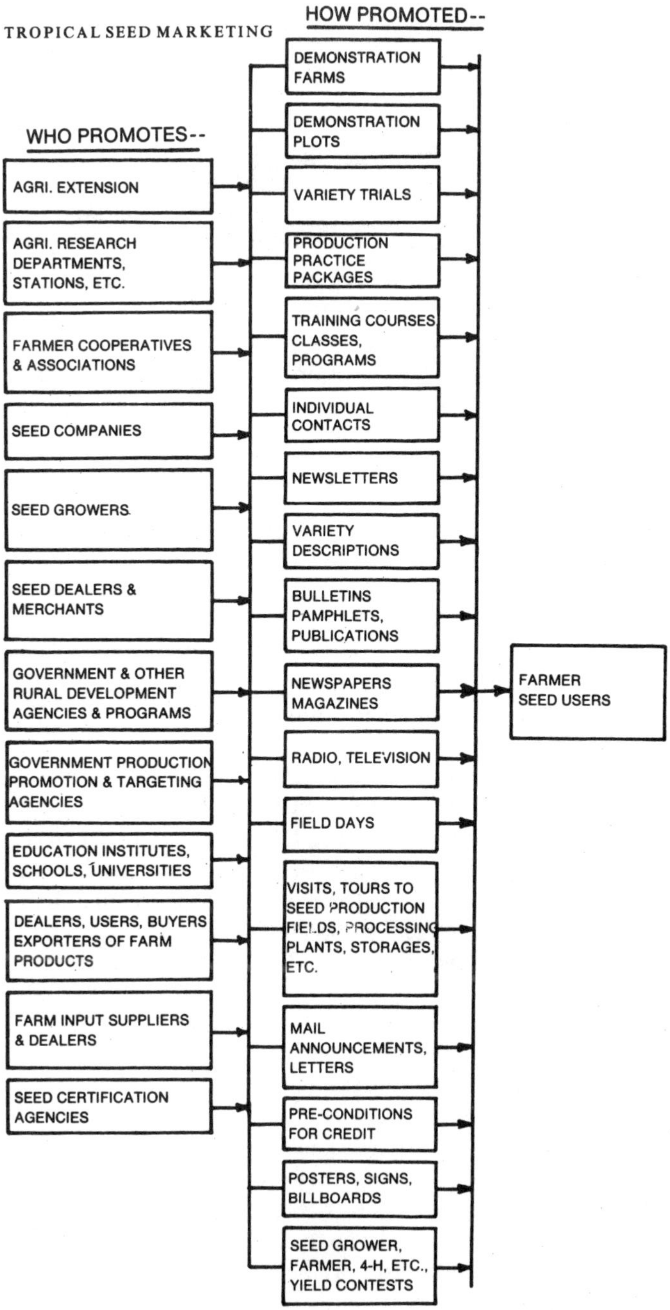

Figure 5. Government and private sector agencies which should participate in coordinated comprehensive promotion of improved seed to small farmers, and promotional methods which have been successfully used.

Distribution

Seed should be produced in areas which offer efficiency and economy in production, and then marketed to farmers in a larger area. Many channels have been successfully used to move seed to farmers.

Seed must be taken to the farmers who need it, at the time and in the amounts needed. Few small farmers will go to any trouble or travel any distance to obtain improved seed; it must be taken to them so that it is almost as available as alternative kinds of seed, and is within their normal 'zone of mobility'.

Distribution is the marketing component which moves seed from the processing centre to the farmer. The sole purpose of the distribution system and its channels is to get seed to farmers. The cost, time, distance and efficiency of moving seed determine the structure of the distribution system. It requires:

1. Adequate transport to move the required seed to the required points, in time for proper planting.
2. Storage which can maintain seed quality, at every point where seed will be held.
3. Stocking points and retail outlets which make seed available relatively near to all target farmers.

Marketing channels

The classic channels of seed marketing are shown in figure 6. Seed move from the processing centre to brokers or wholesalers; smaller amounts then move to retail dealers who supply seed directly to farmers.

The pattern of movement varies with different crop seed; the objective of distribution is to get seed to farmers efficiently; when a different means is effective, it should be used. Figure 7 shows marketing channels used in the USA for peanut (*Arachis hypogaea* L.) seed, usually produced in the area of use. Seed produced outside the area of use, such as forages in the USA, is distributed through different handlers, as shown in figure 8.

Distribution channels for seed produced by the Thailand Seed Development Programme are shown in figure 9.

Transport

Any suitable transport may be used. Rail transport is suited to moving large amounts between wholesale distribution points, if it is not delayed. Trucks can supply almost any amount to any location. Some seed distribution systems send delivery trucks on specific routes at regular intervals during seed-selling seasons, to supply each retail dealer the seed his farmers need. If the dealer has special needs, he advises the supplier by telegram or telephone, or the supplier may telephone the dealers to get their exact needs. To serve small retail dealers in villages, seed may be transported by truck to a wholesaler or large retail dealer located nearby. Seed is then transported to small village dealers by pick-up, car, bullock cart or any available transport, even sometimes by boat.

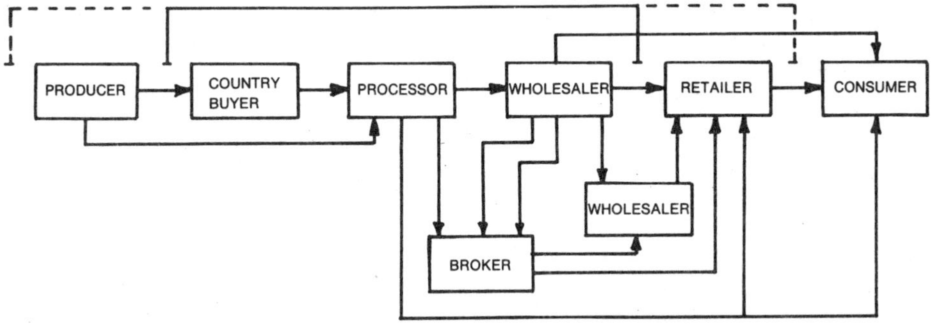

Figure 6. The classic channels of seed marketing, as developed from the US seed industry (after Wheeler and Hill, 1975).

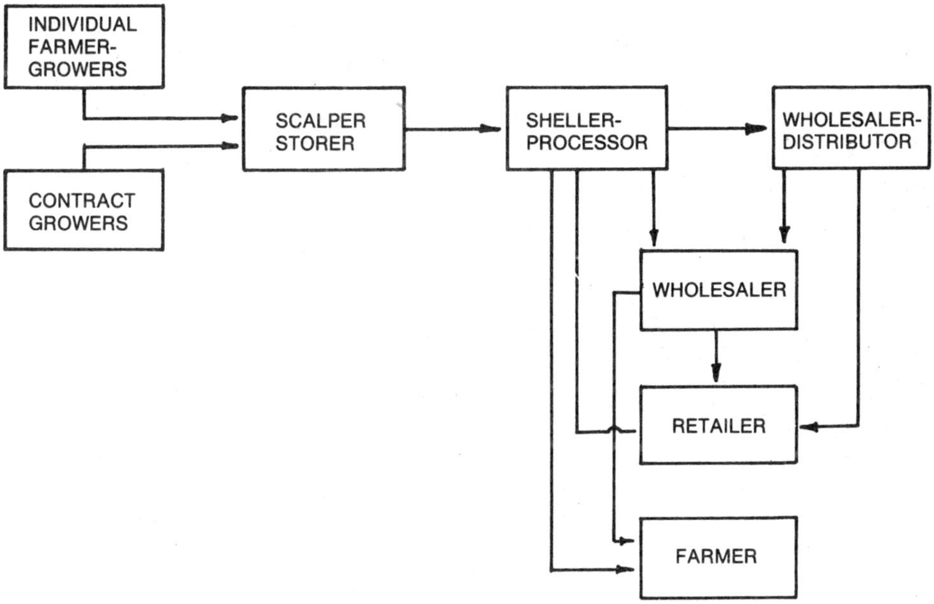

Figure 7. Marketing channels in the USA for distributing peanut (*Arachis hypogaea* L.) seed to farmers (after University of Georgia).

Seed may be transported from the dealer to the farmer by any suitable available means, operated by the dealer, farmer or other person.

The primary requirement of transport is to move the required seed so that it is available to farmers when they need it, and to protect seed quality. During transport,

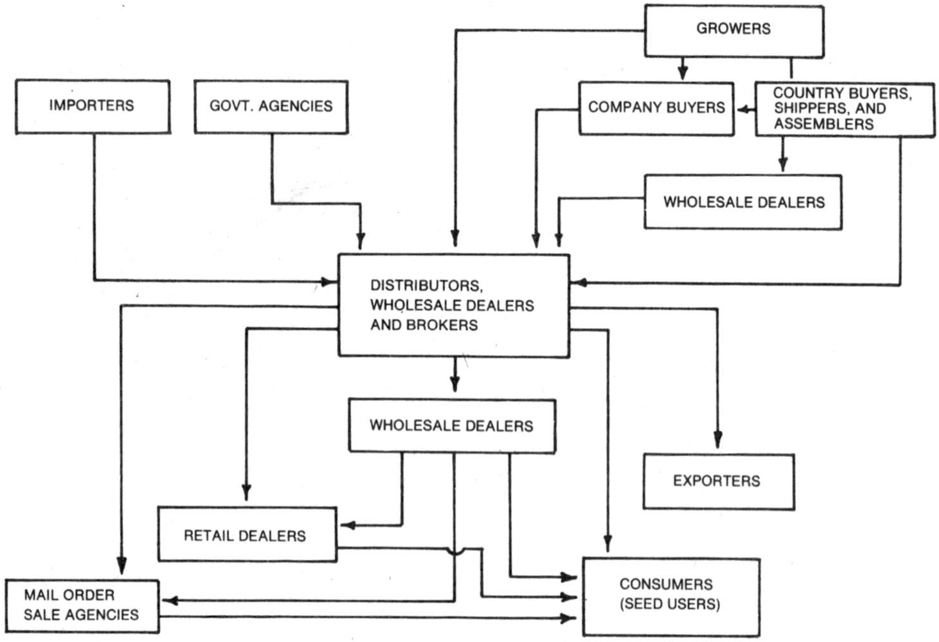

Figure 8. Marketing channels in the USA for forage crop seed produced outside of the region where they are used (after Wheeler and Hill, 1975).

seed is still in storage and safe storage conditions must be maintained (dry, cool, protected from damage by insects/pests) and identity must be maintained. Some tropical seed programmes use excellent watertight, thermally-insulated van trucks to transport seed.

Storage

Seed programmes usually have good storage facilities at the processing centre. However, until the farmer plants it, seed is still in storage and must be kept under safe conditions. The retail seed outlet should have a small special seed storage unit, installed separately or in an existing building, to keep seed dry, cool and protected from insects, pests and damage. A pre-fabricated insulated seed storage room installed inside a building is ideal. A good special retail seed storage not only helps maintain seed quality, but also has promotional value.

Low-moisture seed packed in vapour proof bags maintains its quality during distribution. Very dry seed withstands higher temperatures, so special cooling equipment is not needed during transport or retail storage. Moisture and vapour cannot penetrate into the bag, so seed moisture is kept at safe levels and insect and fungal damage is prevented. Very dry seed, however, is more susceptible to mechanical injury and

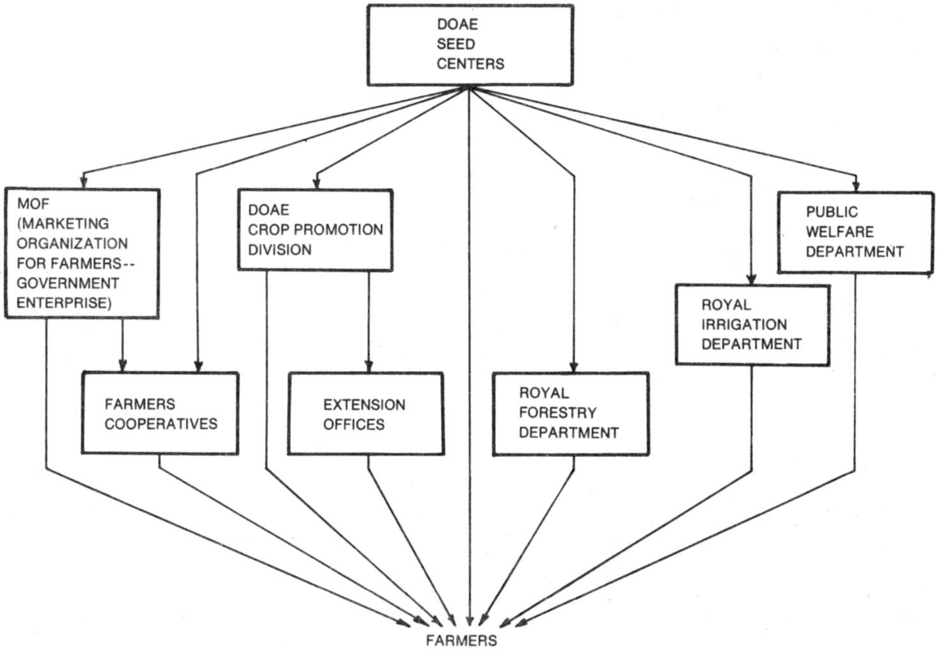

Figure 9. Marketing channels used by the Seed Division, Department of Agricultural Extension (a government seed programme), to distribute improved seed of field crops to small farmers in Thailand.

the bags must be handled gently. Proper vapour proof bags and careful handling also have considerable promotional value.

Retail dealers

Retail outlets sell seed directly to the farmers who plant it. Successful seed marketing requires enough trustworthy, reliable dealers to make improved seed readily available to the target farmers.

Retail dealers are the key to seed marketing; they are where farmers get seed, and they make the actual sales contact and delivery. Retail dealers must not only sell seed, but also be able to tell farmers why improved seed is better, why the recommended cultivars better, how and when to plant it, how to handle and harvest the crop for better yields. The 'point-of-sale' educational promotion by the dealer is the final convincing effort to get the farmer to use improved seed. The retail dealer must hold the farmers' trust and willingness to deal with him.

A major part of seed promotion is intended to establish contacts with farmers so they can be 'sold' improved seed. The best seed dealers are existing merchants, shops, buyers, cooperatives, offices, etc., which already have established contacts with farmers. Using them as seed dealers not only uses existing farmer contacts and helps

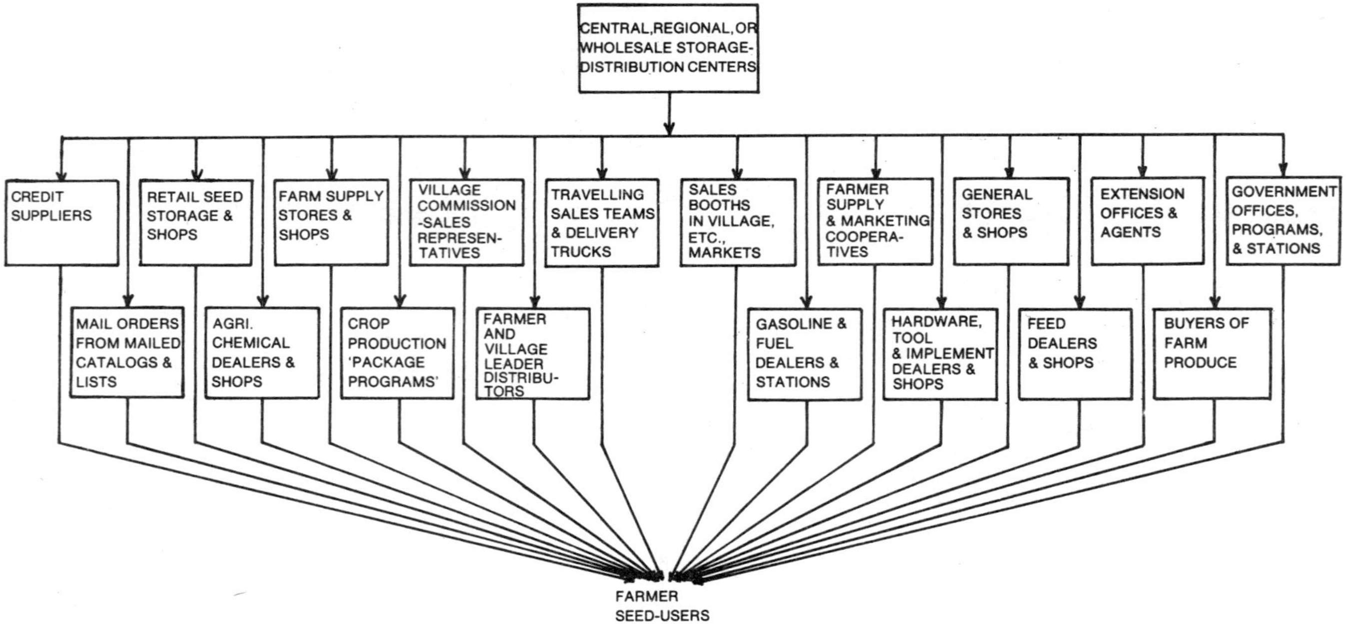

Figure 10. Retail outlets and marketing channels which have been used successfully to market seed to farmers in the tropics.

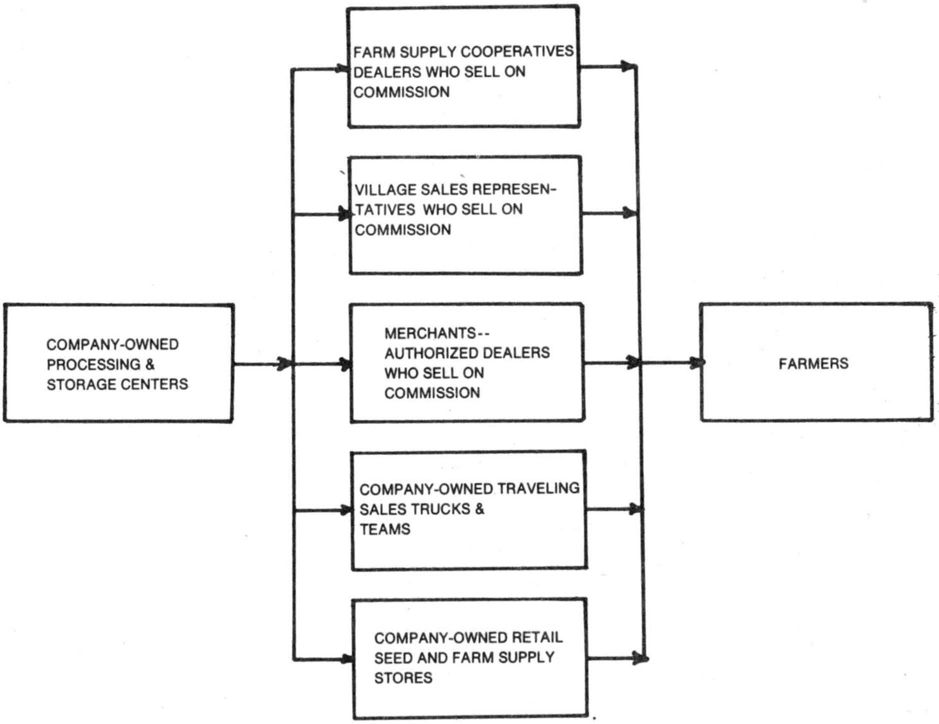

Figure 11. Retail dealer and marketing system used by a successful tropical seed programme.

distribute seed widely, but also gains the merchants' support for the improved seed programme.

Retail outlets which have successfully marketed seed to farmers in the tropics are shown in figure 10; the dealer system used by one tropical programme is shown in figure 11.

Retail seed dealers should be located so that improved seed is available within farmers' normal 'zone of mobility', or the distance they normally travel to buy other supplies, sell their produce, etc. If farmers must travel further to get seed, the number who use it declines as distance increases. Mobility of farmers varies according to road conditions, farmer economic level, availability and cost of transport, location of markets which supply farmers and buy their produce, and other factors.

Retail seed dealers should be conveniently located in every city, village and market which deals with the target farmers. Maximum use of improved seed is achieved when it is available in every village from local seed dealers, sales booths at markets, commission-sales seed representatives, farmer or village leader seed distributors, travelling sales-delivery trucks, local supply shops and other respected outlets.

Retail seed dealers may sell only seed, or may also sell any other goods and services which farmers buy, and may even buy farmers' produce. All seed dealers, regardless of their other operations, must follow practices which ensure long-term trust and good relations with farmer-customers. Both the seed programme and national crop production suffer if the seed dealer cannot get 'repeat sales'--farmers buying and using improved seed year after year.

Effective practices for retail seed dealers under tropical social and economic contexts include:

1. Be ethical, honest, and fair: farmers must trust the seed dealer and be convinced that he is honest, charging them a reasonable price and giving them fair measure. This is the most important requirement to ensure 'repeat sales'.
2. Sell only high-quality seed: farmers must have confidence in the quality of the seed. Test all seed and discard low quality lots--a small loss today is better than losing the entire business next year.
3. Know the seed and cultivar: know, and advise farmers honestly on what improved seed and recommended cultivars can do.
4. Know appropriate farming practices and requirements: advise farmers not only on improved seed and recommended cultivars, but also on better cultural practices, what crops to grow, marketing and other helpful information.
5. Maintain proper storage: prevent loss of seed and seed quality and use good storage to build a good reputation.
6. Keep the store attractive and clean: attract and impress farmer customers with the dedication and competence of management and staff.
7. Keep up-to-date records: to help follow-up farmer-customers, as a guide for future seed sales and to comply with laws and seed programme requirements.
8. Follow up the sale: visit or check with farmer-customers later to see how the improved seed performed and what they will purchase the next season.
9. Use promotion and publicity wisely: stimulate understanding of and demand for improved seed and let farmers know where they can get it.
10. Know the competition: know what alternative sources of seed the farmer has, what their advantages and disadvantages are, and how to sell improved seed effectively against the competition. In a tropical agriculture, the primary competitor is usually low-quality 'farmer seed', not other seed programmes.

The seed programme must support and assist retail dealers with:

1. Adequate stocks of high quality seed, in proper packages, at the right time to meet farmer needs.
2. Necessary credit, or help in obtaining it, to maintain the seed stocks needed to supply all potential farmer customers.
3. Information and data on seed, cultivars, fertilisers, pesticides, cultural practices, etc., which help the farmer.
4. Information on crop and market conditions which affect farmer use of seed.
5. Support, guidance, training materials and cost-sharing for promoting and selling seed.

6. Records, reports and ordering systems to help maintain efficient seed sales and supplies.
7. Information on seed programme policies, procedures and services.

Forecasting sales

Unless farmers plant the improved seed produced, the investment in a seed programme is wasted.

Developing seed programmes often have surplus seed although many farmers do not use improved seed. At the same time, production of some seed kinds may be far below the demand.

Seed production must be according to demand, confirmed and potential, as to amount, kind, cultivar, quality, price, time and place.

In a developing tropical seed industry, the strong promotional effort and higher yield potential in early development stages cause demand for seed by leading farmers to grow faster than supply. This attracts more seed producers and stimulates production, and supply increases rapidly. However, the initial increase in the number of farmers who buy it often does not keep pace with the increasing supply of seed. Lack of adequate quality control systems, some producers' desires for maximum short-term profits at the expense of long-term reputation for quality, lack of adequate market intelligence to guide production, changes in farmer confidence in seed quality, farmers planting their own seed after getting the improved cultivar, changes in markets for farm products, saturating the seed needs of more receptive farmers, failure to promote use of improved seed by less progressive farmers, etc., cause the rate of growth in farmer acceptance to slow down. Without adequate market intelligence to guide both production and marketing, production increases disproportionately, and seed producers suddenly face an artificial surplus situation.

Over-production increases costs due to the extra cost of carryover and seed lost due to deterioration. Under-production is costly to the agricultural-economy since not all farmers can get improved seed, and to the seed marketing system because it cannot reliably meet farmers' demands. For a stable seed programme, production must be based on reliable market intelligence so it matches the real demand, and the marketing system must constantly create new demand.

Accurate sales forecasts are essential to permit accurate decisions on production and marketing. To prevent constant swings from surplus to shortage, production must be market-oriented and integrated with dynamic marketing, with both based on adequate and accurate market intelligence. This is critically important in a developing tropical seed industry.

To produce seed according to what can be sold to farmers, it is necessary to know:

1. What seed, crops and cultivars should be used by farmers?
2. What is the real and potential demand for seed: what kind, how much, when, by whom, where, at what price?

This information should help answer the questions:

1. What seed crops and cultivars should be produced? How much? Where? When? How? At what production costs?
2. To whom can the seed be marketed? Where? When? How? How much? In what package size? At what price? At what promotion and distribution cost?

Market intelligence--reliable information on what farmers will purchase--should include:

1. Total area of each crop cultivated each season, rainfed or irrigated.
2. Relative importance of each crop, both to individual farmers and to the region.
3. Number of farmers growing the crop, the range and average size of field grown by individual farmers, how much farmers will spend to produce the crop and what their income should be from the crop.
4. Crop production problems and methods.
5. How much improved seed of higher-yielding cultivars is already being used, and its sources.
6. Farmer preferences for crops and cultivars; seed package size and kind; seed quality and price.
7. Most effective kinds of publicity and promotion.
8. Prices, sources and amounts of improved seed likely to be available this season.
9. Weather, market, supply, prices etc., likely to affect farmers' planting plans and use of seed.
10. Farmer experience, education, outlook, attitudes, economic status, limitations, problems, willingness to use improved inputs and seed, and the type and intensiveness of educational promotion necessary.
11. Improved cultivars available and what advantages they offer.

Demand forecasts to guide production and marketing should be prepared by a specially-designated section to ensure that it is done completely and realistically, based on reliable 'feedback' from farmers through retail seed dealers, extension programmes, research agencies, market report and forecasts, government targets, credit and price support programmes, farmer cooperatives and organisations, and all possible sources.

References

Gregg, B. R. (1980). *Practical Seed Marketing*. 2nd Association of Southeast Asian Nations Seed Technology Workshop. Bangkok, Thailand.

Law, A. G., Gregg, B. R., Young, P. B., and Chetty, P. R. (1971). *Seed Marketing*. National Seeds Corporation, New Delhi, India.

Wheeler, W. A. and Hill, D. D. (1975). *Grassland Seeds*. D. Van Nostrand, Princeton, New Jersey, USA.

Hanssen, K, B. (1983) *Seed Sci. & Technol.*, **11**, 149–156

The production of certified seed for the tobacco industry

K. B. HANSSEN

Seed Services, P.O. Box 8100, Causeway, Zimbabwe

(Accepted August 1982)

Summary

Seed certification can only achieve its aim if there is seed of high quality to be promoted. Certification alone produces no seed. Therefore, there must be tobacco growers who wish to produce certified seed, and are prepared to comply with the regulations and standards laid down in the certification scheme. The care and attention to detail that goes into the production of certified tobacco seed is the prospective buyer's guarantee of a high quality product. The part the producers must play in preparing the product for sale and, in particular, the recommended practices to be followed in the growing, cleaning and treating of tobacco seed are described.

Résumé

La production de semences certifiées pour l'industrie du tabac

La certification des semences ne peut atteindre son but que s'il y a des semences de haute qualité à promouvoir. La certification seule ne produit pas de semences. Par conséquent, il doit y avoir des producteurs de tabac qui désirent produire des semences certifiées et sont prêts à se conformer aux règlements et normes du système de certification. Le soin et l'attention jusque dans le détail qui interviennent dans la production des semences de tabac certifiées sont pour l'acheteur éventuel la garantie d'un produit de haute qualité. On décrit le rôle que les producteurs doivent jouer dans la préparation du produit pour la vente et, en particulier, les pratiques recommandées à suivre en matière de culture, de nettoyage et de traitement des semences de tabac.

Zusammenfassung

Die Produktion von zertifiziertem Saatgut für die Tabakindustrie

Die Saatgutzertifizierung kann nur ihren Zweck erreichen, wenn für Saatgut von hoher Qualität geworben wird. Die Zertifizierung allein erzeugt noch kein Saatgut. Deshalb muß es Tabakpflanzer geben, welche zertifiziertes Saatgut erzeugen wollen und bereit sind, die im Zertifizierungsschema niedergelegten Verfügungen und Vorschriften zu erfüllen. Die Sorgfalt und Aufmerksamkeit bis ins Einzelne, welche bei der Produktion von zertifiziertem Tabaksaatgut vor sich geht, ist die Garantie eines hochqualifizierten Erzeugnisses für den zukünftigen Käufer. Die Rolle, welche die Erzeuger bei der Vorbereitung des Produkts zum Verkauf spielen müssen und im besonderen die empfohlenen Praktiken bei der Anzucht, Reinigung und Behandlung von Tabaksaatgut werden beschrieben.

The Concise Oxford Dictionary definition of quality is 'degree of excellence, relative nature or kind or character'. Within the seed industry, quality, by regular usage, has come to have a very specific meaning related to describing the number of properties which are inspected or investigated. For commercial or standard grade, seed is usually required to meet minimum standards in analytical purity and germination percentage and these properties are subject to routine seed testing in order to ensure that the purchaser receives a product of high physical purity and good germination capacity. For certified seed, two other factors, trueness-to-type and freedom from disease are also determined; field inspections of the plants producing the seed are necessary to investigate these factors. Thus, only a seed lot reaching a satisfactory level in all these properties should qualify as high quality seed.

Seed quality control, however, is unfortunately restricted to certification-type procedures and 'routine seed testing'. Although these are excellent controls in themselves they are passive rather than active. They do not involve positive action to ensure timely harvesting, control of pests and diseases and measures to maintain or upgrade viability.

Introduction

Before a detailed account of current field practices for the production of tobacco (*Nicotiana tabacum* L.) seed is given, a brief discussion of the control measures that are administered to ensure the production of high quality seed is described.

The stock seed to plant the seed crop must be obtained direct from the originating authority and this seed, called breeders' seed, is released each year to members of a seed association who have registered to grow crops. Each seed-grower may register only one cultivar and is permitted to grow only one seed crop from breeders' seed. During the production of the certified seed, at least four field inspections are carried out and at each visit the crop must comply with the field standards. The marketing of the certified seed is the responsibility of the seed association and therefore, before sale, it must submit a sample for routine seed testing.

At the seedbed

For the seed producer, the season starts with the sterilisation and fumigation of an individually fencedoff seedbed isolated by at least 50 m from the commercial seedbeds. Between August 15 and September 15 this bed is seeded with breeders' seed, aiming for a population density of 400–500 seedlings per m^2. The seed must be carefully sown to obtain uniformly spaced plants; too thickly sown beds will produce very spindly plants and too widely spaced plants will result in 'cabbage-like' seedlings. A desirable seedling for transplanting should be 15 cm long by 7 cm thick with 8–10 leaves.

Cleanliness

Both seedbed and seed crop are managed in the same way as the commercial crop but extreme care is taken in personal hygiene. The individual in charge of production should be a nonsmoker (or non-snuffer or chewer of tobacco) and he must also have facilities to wash his hands and feet before entering the seedbed site. During transplanting, workers are also required to take the same precautions before they start to plant out.

Plot selection and planting

If the seedlings pass the first field inspection, they are transplanted from the seedbed into a field plot. This must be at least 400 m from tobacco of the same or any other type and far enough away from buildings which have housed tobacco leaf in any form. The producer must take into account not only his own fields but also those of his neighbours. It is usual to mark the equipment used in the seed production unit with a large 'S' and keep it separate from the equipment used for the commercial leaf crop.

Priming and spraying

Once the tobacco seed plot is planted, extra care is taken to prevent disease and insect pests. This involves an intensive spraying programme: twice a week for anthracnose (*Colletotrichum tabacum* Böning), at five day intervals for frogeye, (*Cercospera nicotianae* Ell. & Ev.), wildfire (*Pseudomonas tabaci* (Wolf & Foster) Stevens) and angular leaf spot *P. angulata* (Fromme & Murray) Stevens) and once a week for aphids and other pests. This programme is followed from start to finish, i.e. seedlings to ripe seed.

Tobacco seed was first produced under certification in 1958, because of the appearance, in Zimbabwe, of anthracnose, a serious seedborne disease, and the associated danger of unrestricted spread in the absence of seed production controls. The second inspection takes place within three weeks of establishing the field plot and before the plants drop their seedbed leaves. Thus, at this inspection, it is again possible to examine all the leaves and particularly the bottom ones for anthracnose. As an additional safeguard against the risk of disease infection from the soil, the producer must now comply with the following rule: *Soon after the second field inspection, starting with the lowermost seedbed leaves, the leaves of the lower third of the plant shall be removed progressively, irrespective of degree of ripeness. By the time flowering begins, this operation shall be complete, leaving a distinct free space between the remaining leaves and the ground. When the plants come into full flower the top five leaves shall be removed.*

The main object of these precautions is to eliminate the possibility of potentially dangerous seed-borne diseases spreading from the soil to the base of the plants and thereafter, up the stem to the seedhead.

Roguing

Under certain conditions some cultivars are prone to produce off-type plants. The frequency of such off-types is seldom more that two to four plants per hectare. Two kinds of off-type plants are usually observed. One has a fairly typical broadleaf shape and the lateral veins of the leaves meet the midribs almost at right angles; such plants are readily recognised shortly after transplanting. The other off-type generally has normal leaves on the lower and central stalk positions but, on the upper portion of the plant, the leaves are intermediate between normal and broadleaf. In most cases such plants can only be detected shortly before flowering.

These off-type characteristics may be transmitted to the normal plants in the plot and therefore every offtype, whether suspect or real, must be topped before it has any open flowers. This means that it is essential to ensure a uniform stand of plants so that any offtypes can be immediately rogued from the field. It is normal to limit field plots to a fixed plant population (8,650 plants) in order to comply with the following rule.

The area planted may be 25 percent greater than the area designated by the Association when the crop is registered. Immediately after the second field inspection, and at regular intervals thereafter until the final field inspection, each member is expected to inspect his crop thoroughly and to top (and sucker) all undesirable plants. At flower bud stage, a compulsory minimum of 20 per cent of the plants must have been topped, but if the member so desires he may top as high a proportion of plants as he sees fit over and above the prescribed minimum. Compact areas of poor growth such as may occur in water-logged portions of the field, must be excluded from the seed production plot proper and shall not form part of the compulsory 20% portion that must be topped. The aim of this rule is to ensure that the producer makes regular visits to the seed plot in order to ensure the field has uniformity of growth at the time of each inspection.

Harvesting

Having successfully surmounted the hurdle of the final field inspection, the producer of the tobacco seed should harvest the crop when sufficient seedheads are ripe. Reaping of the seedheads takes place when the seed capsules are brown and the seed is comparatively dry and olive-brown in colour; white seed is immature. Because all capsules on the head do not ripen simultaneously, reaping may have to be repeated more than once. In a normal, reasonably uniform crop about three harvestings should remove all the seedheads. Any immature seedhead at the final harvesting should be abandoned.

Drying and threshing

Only one cultivar may be handled in any one season. Where the possibility of mixing cultivars exists, all trace of the previous cultivar (including packets for sale), must be removed from the cleaning premises before the new seed crop is delivered for processing.

Once seedheads are collected they may be dried in a specially prepared barn with artificial heat or air dried in a shed which has never contained tobacco in any form. Before it is filled, the barn must be plastered throughout and sterilised by heating to 82 °C for a minimum of four hours. The seedheads may be placed on netting wire stretched across the tier poles or tied in bunches of five or six seedheads, with a tobacco stick threaded through about six of the bundles, suspended normally on the tier poles. The floor should be covered with tobacco baling paper or polythene and any seed falling outside this cover during drying is discarded.

When the seed capsules are 'tinder-dry' they should be removed from their bundles and crushed or threshed through a 6.27 mm round-hole sieve. In this way the large fragments of stalk and capsule are removed. Crushing is usually done by hand. The crushed material is then passed over suitable screens to sift out impurities such as large sand particles and remove the remaining larger fragments of the pods. A poor job of cleaning can often be attributed to a lack of proper sieve.

Cleaning

(a) *Sieving*

Usually mosquito gauze or a No. 14 (1.2 mm) mesh sieve is suitable for the first cleaning and then a very fine No. 30 (0.5 mm) mesh sieve is used to remove the remaining larger fragments of the pods. It is important to note that any fragments which weigh as much as tobacco seed must be removed by sieving. It is clear, therefore, that the final sieving should only just allow the seed to penetrate. Any trash that is the same size as the seed will not be as heavy and this can then be readily removed by blowing through a specially designed tobacco seed blower.

(b) *Blowing*

The air blast cleaner (figure 1) consists of a hopper, chute, an open tunnel, three collecting bins and an electric fan. The efficiency of the separating depends upon the velocity of the air current and the rate at which the material is fed into it. Adjustments are provided to control the velocity of the airstream as well as the rate of feeding. Adjustments to the velocity of the air flow are made by varying the distance between the fan and the feed point.

The operation of the airblast cleaner is quite simple. The light and heavy seed mixture falls from the hopper on to a chute and then through the air stream which removes the light material. The rate of feed must be controlled to permit a uniform layer of seed to be evenly distributed throughout the width of the gate of the hopper. Similarly, the volume of seed falling onto the chute must also be spread evenly. Overloading the chute will cause a large amount of light weight material to be carried off by the heavier seed. The position of the fan is also important; too little air will result in too much trash while too much air will blow off much of the good seed. It is generally accepted that about 50% of the volume of seed through the cleaner should be discarded, if heavy plump seed of the highest germination percentage is to be packed and offered for sale.

Chemical seed treatment

After cleaning, the seed producer must treat the seed with a solution of silver nitrate to eliminate any disease organisms such as wildfire and angular leaf spot that may have adhered to it. This is a delicate operation for silver nitrate is extremely toxic and will impair the germinating capacity of the seed if the concentration is too high or contact time with the silver nitrate solution is too long. The rate at which the seed is dried is also critical, slow drying being detrimental. This treatment is usually conducted out of doors during spells of dry sunny weather. Producers who treat the seed indoors use an alternative method. The procedure is as follows:

1. A calm, sunny day is chosen, alternatively a clean room is prepared.

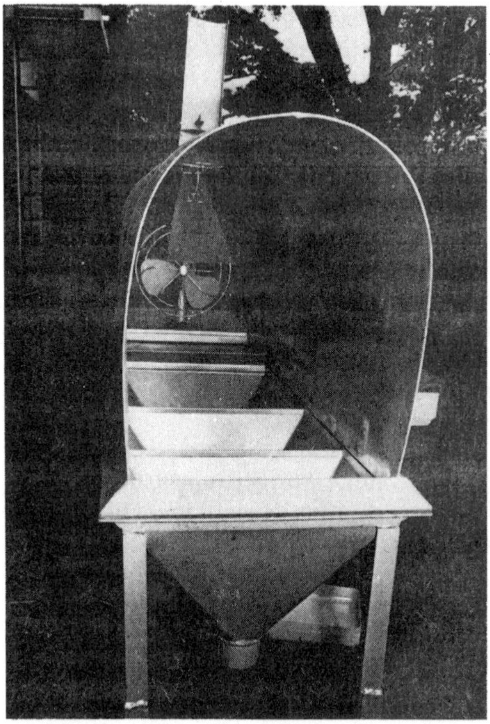

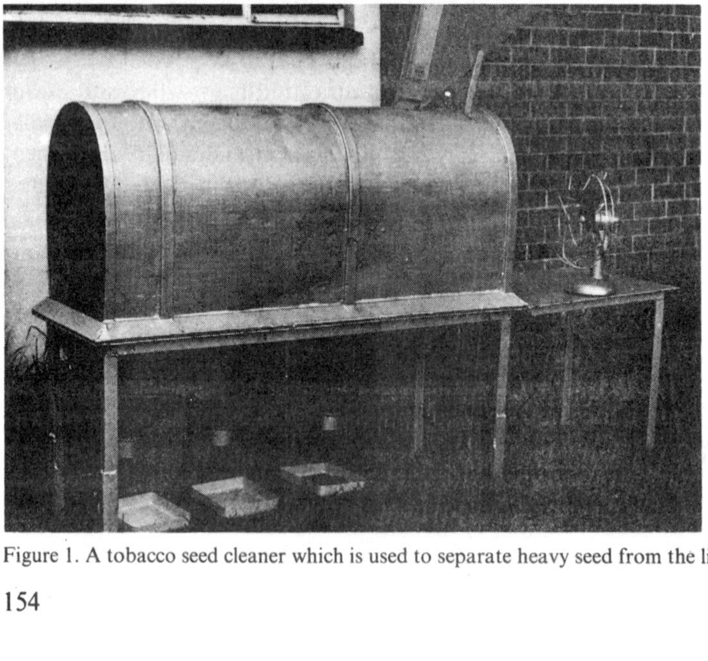

Figure 1. A tobacco seed cleaner which is used to separate heavy seed from the lighter factions.

2. Approximately $2\frac{1}{2}$ kg of seed are measured on to unbleached calico cloth of about 20 × 24 cm size and tied into the shape of a bag. Large seed lots require too much time to dry and the result is poor germination. A calico of 24 threads per cm is satisfactory, alternatively, only 0.5 kg of seed is measured on to calico at any one time for the indoor treatment.
3. Measure the appropriate quantity of water to be used for making up the solution into a plastic bucket. Allow 17.5 l for each parcel of 2.5 kg of seed.
4. Dissolve sufficient silver nitrate to give a 1:1000 concentration, i.e. 1 g of silver nitrate to 1 l of water.

   The silver nitrate which is obtained in tablet or granular form should be stored in the dark or in brown glass bottles to prevent decomposition.
5. Immerse the calico bag in the solution for 15 minutes and squeeze the contents at five minute intervals to ensure that the solution reaches all the seed. Protective gloves should be worn.
6. After 15 minutes, the bag is removed and squeezed repeatedly to expel the excess solution. The bag is opened and the seed spread out on the cloth to dry in the sun. The temperature on the surface of the seed should not exceed 38 °C. Although temperatures of up to 50 °C have allegedly not impaired germination, they are not recommended. The seed should be stirred regularly to hasten the drying time which should not exceed two hours. Another method for drying large quantities of seed during wet weather is to stretch the calico sheet over a tray with an 18 mm mesh chicken wire bottom. The tray is placed on a stand which ensures a 300 mm clearance above the barn flues when placed over them. The barn flues are heated slightly and then an electric fan is used to assist in air circulation. In this way up to 30 kg of seed can be treated per barn per day. Alternatively, for the indoor method, the calico sheet is stretched over a tray with a chicken wire bottom and the seed is dried from the heat emanating from two 150 w electric light bulbs which are set 0.6 m above the table. Do not let the thermometer reading exceed 38 °C. Keep stirring continuously throughout the drying time which again should not exceed two hours.
7. Before packing, the seed is left to air dry in shallow basins for seven days and stirred at regular intervals.

Conclusion

The activities required to be undertaken by tobacco seed producers have been outlined. These activities are made effective by field inspections of the plants growing in the field and testing of the seed offered for sale. However, it will be seen from the above account of the field practices that, it is the producers themselves who are mainly responsible for the supply of good quality seed to the industry. Thus, as in any area of production, a large share of the credit belongs rightly to the producer. Tobacco seed production is no exception to the rule.

Acknowledgement

Appreciation is expressed to the members of the Zimbabwe Tobacco Seed Association and to the staff of the Tobacco Research Board of Zimbabwe who have freely contributed information concerning their seed production methods.

Van der Burg, W. J., Bekendam J., van Geffen, A. and Heuver, M. (1983), *Seed Sci. & Technol.*, **11**, 157–227

Project Seed Laboratory 2000–5000

(Second, revised edition)

W. J. VAN DER BURG[1], J. BEKENDAM, A. VAN GEFFEN and M. HEUVER[2]

Government Seed Testing Station, P.O. Box 9104, 6700 HE Wageningen, The Netherlands

Contents

Summary

A description is given of how a seed testing station in tropical or subtropical areas could be established. Two alternatives are described: Seedlab 2000, that can test about 2000 samples per year and Seedlab 5000 that can test at least 5000 samples per year. Directives and general considerations concerning the staffing, the organisation of the work, the lay-out of the building, and the equipment needed are given. Forty-six figures and two tables give an impression of the equipment and administrative forms used. Equipment which in the experience of the authors has been found suitable for the work is recommended, and detailed descriptions and company addresses are mentioned. Some equipment that is not commercially available is described and plans for construction are included. A list of books and journals for a basic seed testing library is appended.

[1] Editor

[2] Present address: State Institute for Quality Control of Agricultural Products, P.O. Box 230, 6700 AE Wageningen, The Netherlands.

Résumé

Projet de laboratoire de semences 2000–5000

On décrit comment une station d'essais de semences peut être établie dans les régions tropicales ou subtropicales. On décrit deux propositions: 'Seedlab 2000' qui peut analyser environ 2000 échantillons par an et 'Seedlab 5000' qui peut analyser au moins 5000 échantillons par an. On donne les principes directeurs et les considérations générales concernant le personnel, l'organisation du travail, la disposition du bâtiment et l'équipement nécessaires. 46 figures et deux tableaux donnent une idée de l'équipement et des documents administratifs utilisés. On recommande les équipements, qui d'après l'expérience des auteurs se sont révélés adaptés au travail et on fait état des descriptions détaillées et des adresses de compagnies. On décrit quelques matériels qui ne sont pas disponibles dans le commerce et leurs plans de construction sont inclus. Une liste de livres et de périodiques pour une bibliothèque de base en matière d'essais de semences est donnée en appendice.

Zusammenfassung

Das Projekt Saatgutlabor 2000–5000

Es wird beschrieben, wie eine Saatgutprüfstelle in tropischen oder subtropischen Gebieten errichtet werden könnte. Dabei werden zwei Alternativen angeführt: Das Saatgutlabor 2000, welches ungefähr 2000 Proben pro Jahr prüfen kann, und das Saatgutlabor 5000, welches mindestens 5000 Proben pro Jahr untersuchen kann. Hinweise werden gegeben und Überlegungen angestellt hinsichtlich der Mitarbeiter, der Arbeitsorganisation, der Gebäudeplanung und der erforderlichen Geräteausrüstung. 46 Abbildungen und 2 Tabellen vermitteln einen Eindruck von der Einrichtung und den angewandten Verwaltungsformen. Ausrüstungsstücke, welche nach der Erfahrung der Autoren sich für die Arbeit als brachbar erwiesen haben, werden empfohlen sowie genaue Beschreibungen und Firmenanschriften aufgeführt. Geräte, welche auf Handelsbasis nicht erhältlich sind, werden beschrieben und Pläne für die Herstellung beigegeben. Als Anhang findet sich ein Verzeichnis von Büchern und Zeitschriften, die den Grundstock einer Bibliothek für eine Saatgutprüfstelle bilden.

1. Introduction

In 1979 the Government Seed Testing Station, Wageningen, Netherlands, published a paper called *Project Seed Laboratory 2000–5000* by M. Heuver, J. Bekendam, W. J. van der Burg and A. van Geffen, based on *Project Seed Laboratory 5000* (Proc. Int. Seed Test. Ass. 34 (1), 1969). It consisted of a text part, a binder with technical drawings and a binder with brochures of equipment. One hundred copies were distributed by FAO and another 100 copies by the Wageningen station. It described how a seed laboratory could be constructed, equipped, organised and administered. Alternatives were given for a laboratory that could test about 2000 samples per year and one that could handle at least 5000 samples per year. Certain equipment was recommended, based on the experience of the authors with it. Equipment produced by other manufacturers and used in other countries may however be as good. In this connection readers are advised to consult the Survey of Equipment and Supplies for Seed Testing (ISTA, 1981).

The need was felt to make it more widely available through a concise re-edition in Seed Science and Technology. To this end the information on equipment is updated, much of the information from the binders is included in the text and figures, and recent experience of the authors in developing countries is included. However, one aspect is left out: seed health testing. The original publication gave only very limited information on the aspects of seed health testing and the equipment and training needed, and it was felt that the topic needed a paper of its own to be complete. Besides, the authors believe that seed laboratories of this kind, often located in countries with a developing seed industry, should start by investigating the basic quality aspects of the seed first: moisture, purity, germination.

The authors gratefully acknowledge the assistance of the following persons who have given their support either during the preparation of the original paper or this re-edition (in alphabetical order):
Dr. C. Anselme, C. Bense, Dr. W. P. Feistritzer, Prof. E. E. Hardin, Dr. M. J. Hill, Prof. L. Kåhre, D. B. MacKay, Dr. E. Madsen, Ing. A. Stuurman, J. H. B. Tonkin, C. Witte.

2. Staffing

2.1. General

Seed testing requires an input of manpower that differs for each type of seed. A range of species therefore has to be allowed for in determining staff requirements. The presented estimate aims at the average requirements of a station in a tropical or subtropical region, dealing with 2000 or 5000 samples annually.

The work should be arranged in such a way that germination tests on all samples can be commenced within a reasonable time (24–48 hours) from receipt at the station. Intake of samples is rarely constant, neither in rate, nor in composition. For the laboratory testing 5000 samples annually a peak season lasting possibly three to five months can be expected, with an intake of 1750–3000 samples during that period. The station is expected to handle this peak expeditiously, with no more than a slight delay.

An example of possible staffing is given below. The number of analysts depends very much on the intake of samples in the peak period, the type of seed, the kind and number of analyses required and the experience of the staff.

2.2. Purity analysts

An intake of 1750 samples in three months, each providing 21 working days, gives 28 samples per day. This means that for purity, each sample being subjected to duplicate analysis (two half working samples), 56 determinations per day have to be made. The time needed will then be:

number of determinations	kind of seed	man hours
14 × 2	cereals, rice and maize	9
7 × 2	pulses	4
7 × 2	forage crops, including 'difficult' grasses	19
TOTAL 56		32

Time must also be allowed for performing additional determinations whenever duplicate analyses are not sufficiently in agreement (this is checked by using the tolerance tables provided in the ISTA Rules).

If one analyst provides eight working hours a day, the purity section would be adequately staffed with four trained analysts. Staffing at this rate may sometimes lead to a backlog accumulating, but delays should not be alarming. Overstaffing is felt especially in the slack season as trained analysts cannot normally be laid off then.

One of the analysts should be delegated as responsible for the purity section.

2.3. Germination analysts

An intake of 28 samples per day means that 28 germination tests have to be carried out daily (not including a few tests that will appear not to be satisfactory for one reason or another and therefore have to be repeated). Although the amount of work involved depends on the kind of seed, variations are not as large as in purity analysis.

The work comprises:

- counting and planting 4 × 100 seeds, the substrate being moist paper, sand or soil;
- providing controlled environmental conditions for the tests;
- evaluating seedlings and results; most samples need several consecutive counts before the final result is obtained.

Three trained analysts should be able to handle 28 tests a day.

To supervise germination testing, including the flow of analysis, check on use and accuracy of equipment and maintain supplies one of the analysts should be placed in charge of the germination section.

2.4. Moisture analysts & administrative staff

All seven analysts should be trained in both aspects of analysis: purity and germination. One or more should also be trained to perform moisture determinations.

Unless calculation is carried out by an administrative clerk at least one analyst in each of the purity and the germination section should be trained to handle the calculation of test results.

In addition a seed laboratory needs administrative staff. This is even more so when an established laboratory also deals with seed certification, an activity that involves a considerable administrative burden. A starting seed laboratory however, should have at least one clerk/typist.

2.5. Person-in-charge, training of staff

Finally there will be a person in overall charge. It is essential that this person should have a good scientific background, preferably in botany or agronomy. He also needs at least a three months stay in another station with facilities and experience to provide him with adequate training. The analyst in charge of the purity or germination section should preferably have sufficient secondary education, if possible including professional education in agriculture. He or she should also receive training in another laboratory and attend international workshops from time to time. Such workshops are organised by or in co-operation with international organisation like FAO and ISTA and individual countries (e.g. the FAO-Norway Workshops and regional workshops as organised by the Seed Technology Centre of New Zealand, etc.). The analysts of the laboratory should receive training for an initial period of for instance two years which may be concluded with examinations.

2.6. Total staff required

To summarise, the 5000 laboratory should have the following staff:

1 superintendent
1 clerk/typist
7 analysts for moisture, purity and germination

---

9 persons in total

A smaller laboratory testing 2000 samples annually needs less staff. Four analysts qualified in both purity and germination testing could cope with this intake, one of them being given general responsibility for seeing that the work is done promptly and in accordance with the regulations.

3. The work

3.1. Receipt and registration of submitted samples

A submitted sample is a quantity of seed furnishing sufficient seed for the tests required. It is sent in a linen bag or other container. The submitted sample must be representative of the lot, containing the same constituents in the same proportion. How a representative sample is drawn from the lot and reduced to submitted sample size is prescribed in the ISTA Rules. On receipt, a clerk records particulars of the sample and tests requested, paying particular attention to any special requests.

The use of preprinted forms is recommended for entering the details to ensure accuracy and speed in issue of information by the station. For further detail see chapter 13.

3.2. Purity test

The object of the purity analysis is to determine:

– the composition by weight of the sample and by inference the composition of the

seed lot;
- the identity of the various species of seeds and other material in the sample.

The ISTA Rules prescribe:
- the size of a submitted sample;
- the size of a purity working sample, and how it should be taken from a submitted sample;
- how to separate a working sample into: pure seed
  other seed
  inert matter;
- how to evaluate test results, and how to gather additional evidence if needed;
- how to express and report the test results.

Moreover, the Rules recommend certain technical aids. If used with skill, these aids (see chapter 8) will reduce the time taken by purity analyses.

The purity test also provides the pure seed, which is essential for the germination test.

3.3. Germination test

A germination test provides information with respect to the field planting value of the seed. The ISTA Rules prescribe:
- number of seeds to be counted at random from the pure seed fraction;
- qualities and use of substrata: paper, sand and soil;
- directions for planting the seed;
- treatments for breaking dormancy;
- conditions of moisture, temperature and light that are optimal for growth and evaluation of the seedling;
- definitions by which seedlings are evaluated and rules for interpreting and reporting test results.

The Rules also recommend certain technical aids (see also chapter 9).

3.4. Moisture test

A moisture test provides information regarding the moisture content of the seed. The moisture content is a factor of economic importance because:
- too high a moisture content may result in loss of quality because of mould growth, ageing and increased insect damage;
- the proportion of water being paid for.

A sample of seed for a moisture test must be packed in a moisture proof container so that the sample cannot either dry out or absorb moisture between sampling and testing. In the ISTA Rules test methods are prescribed for each kind of seed.

3.5. Various other items as prescribed and/or recommended in the ISTA Rules

Purity tests for which dry and clean surroundings are essential, and moisture tests that would be badly affected by too high a relative humidity in the atmosphere, should be carried out away from the germination work, preferably in separate rooms.

Such rooms are provided in Seedlab 5000 (figure 1); Seedlab 2000 however has only separate germination and purity benches (figure 4). (Germination tests invariably generate much refuse in discarded seedlings, paper, sand, as well as water spilled on the working tables).

Other determinations which are usually carried out in the purity section are:
- number of weed and other crop seeds present, usually determined in 10 times the quantity of seed that is specified for purity; the seed acts of many countries specify maximum standards for weed and other crop seeds above which seed may not be imported;
- weight of 1000 seeds; this serves as a basis for estimating the weight needed to sow an area of land and can be indicative of the quality of the seed.

Other determinations which may be carried out in the germination section are:
- viability by means of the topographical tetrazolium test (see chapter 10).

The determination of genuineness of species or cultivar may be of interest. Techniques recommended in the ISTA Rules involve the examination of seeds and seedlings in the laboratory and plants in the greenhouse and in field plots.
The Rules also include a chapter dealing with International Certificates, reporting and tolerances. This is of particular interest to official laboratories that intend to join ISTA, and to issue International ISTA Certificates. Information on how to become an accredited ISTA station can be obtained through the ISTA Secretariat.

4. The building

4.1. General requirements

In order to design a building that would provide the facilities for a simple laboratory for routine investigation of seed samples in developing countries, the special requirements of seed testing have been taken into consideration as well as the possible climatic conditions.

The following lay-out requirements have been considered:
1. a maximum of working space in a building of limited size;
2. an efficient and functional lay-out: the movement of samples through the laboratory should be logical;
3. central supervision;
4. adequate illumination and ventilation;
5. water supply and sewage concentrated to minimise plumbing work;
6. ease of extention of the building: allowance should be made for example for extending the administrative section of the building at a future date at moderate cost and effort and without seriously interfering with the structure of the building.

The points above influence both the lay-out of the building and the placing of furnishings and equipment within.

a

b

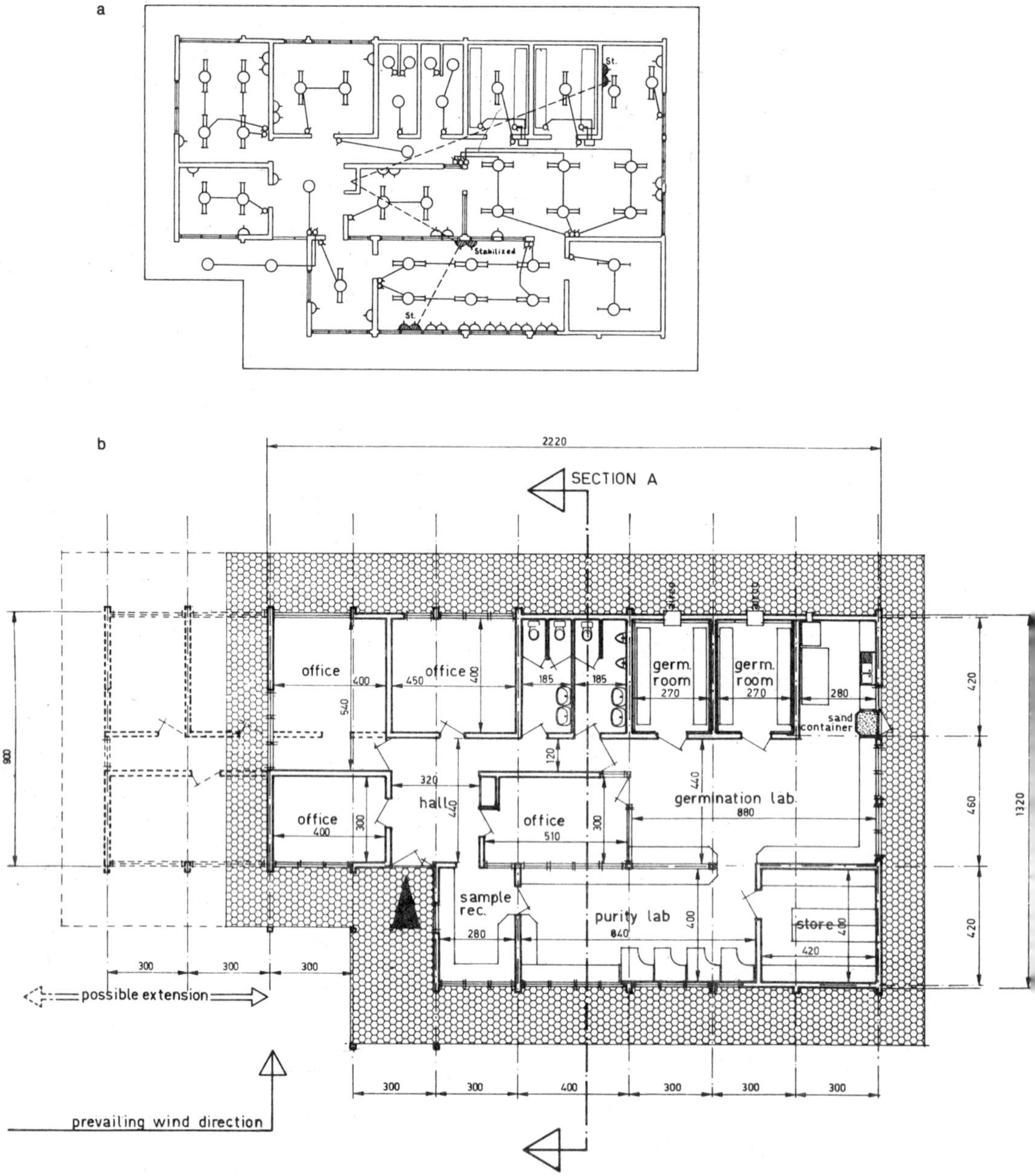

Figure 1. Seedlab 5000, wiring scheme and floor plan.
a. Electrical wiring scheme; dotted lines and hatched power outlets indicate stabilised power supply; b. floor plan, showing the possible extension that can be made, either during building or at a later stage; section A is presented in figure 2. Dimensions in cm.

a

b

c

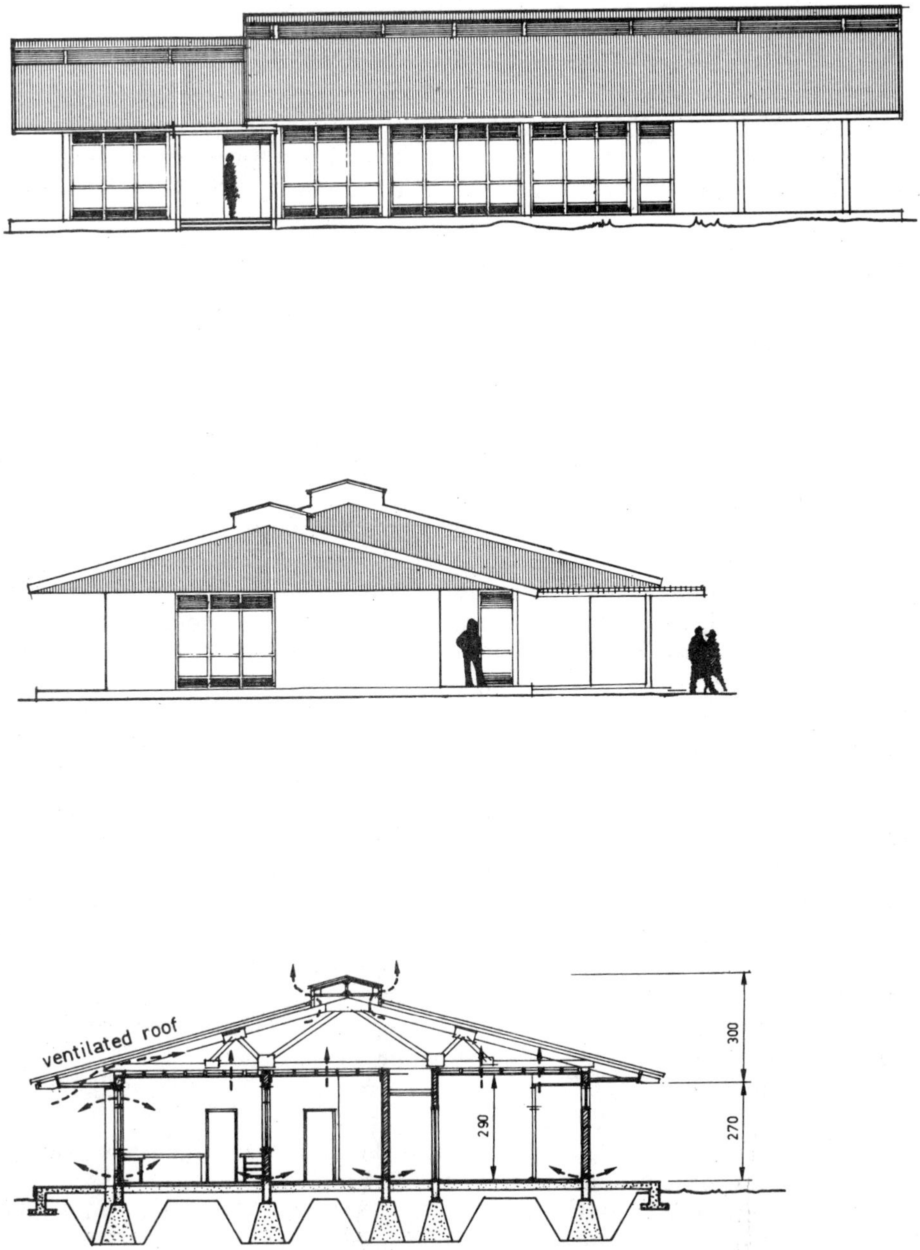

Figure 2. Seedlab 5000, elevations and section.
a. Front elevation; b. left elevation; c. section A as indicated in figure 1.b., showing the provisions for ventilation. Dimensions in cm.

4.2. Lay-out

The building is roughly divided into a laboratory section and an administrative section. If administrative activities increase or expand, the building can be adapted by enlarging it as indicated in figures 1 and 4.

The building has an overhanging roof providing shelter against direct sunlight and heavy rains; in addition it forces the rain water to run freely from the roof leaving a dry gallery around the building.

Entrance to the building is by a central hall, which is surrounded by offices and the sample reception area. From here samples are conveyed to the purity section. In order to make precise weighings a special vibration free weighing table is installed in the purity section. A voltage stabilised electrical supply may be essential for operating the sensitive balances and blowers. A long continuous working bench is fixed to one side with the purity tables opposite, facing the window for good illumination.

A storage room is provided with easy access from both purity and germination sections. The seed must be stored cool and dry and kept free from insects. The maximum temperature should be 15 °–18 °C, so care must be taken in warm and humid climates to avoid condensation on the walls of the storage room (for example by adding insulation material to the walls). Chapter 9 gives details of wall construction.

In the germination section continuous working benches are provided. The space underneath these benches must be kept as free as possible, e.g. storage shelves underneath should be restricted and should not be in the way of analysts working at the bench. Cabinets on turnable wheels that fit under the bench are very convenient in this respect. In one corner there is a sink with an electric water heater (boiler) and an open brick container for sand, height about 50 cm and accessible from the outside for refilling. An electric concrete mixer can also be placed in this area for preparation of sand substrates. A voltage stabilised electrical supply may be essential and supply of clean water for cooling of the germination table and cabinet should be provided. If a cabinet or table with an air-cooled cooling system is chosen a ventilation opening in the wall is necessary.

An aluminium rack system is used in the germination rooms. This can be made following figure 20 or purchased ready made (see chapter 9). Every section has its own upright tubular fluorescent light which can be switched on individually from the front. The whole system is controlled by a programmable time-switch fitted on the outside of the germination room. For insulation directions see chapter 9.

The insulated doors may be bought from a firm specialising in this field, but could possibly be made locally of e.g. non-shrinking hardwood.

The supervisors' office is located centrally in the laboratory, with windows on three sides to give easy observation of all activities in the laboratory.

The office section is grouped around the hall and can also be used to accommodate visitors, or used as storeroom for stationery and equipment.

a

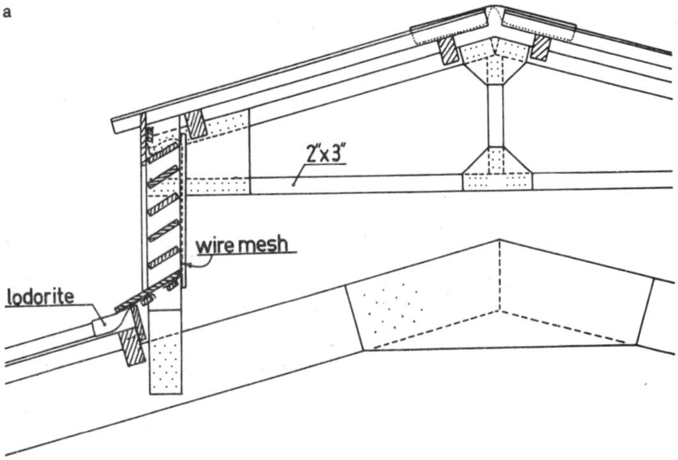

b

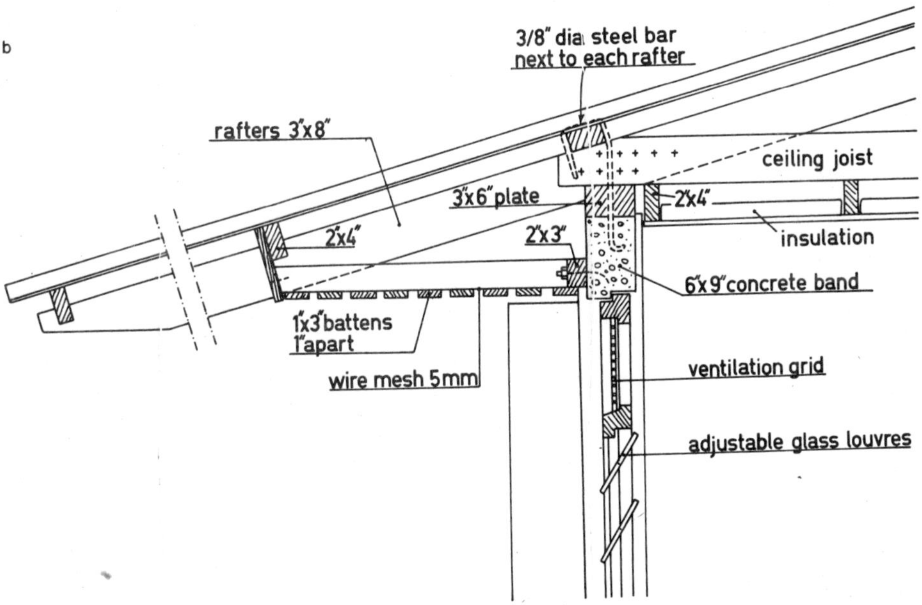

Figure 3. Seedlab 5000, roof construction.
a. Details of ventilation opening on top of roof; b. details of edge of roof showing ventilation opening with wire mesh. All dimensions in inches unless mentioned otherwise.

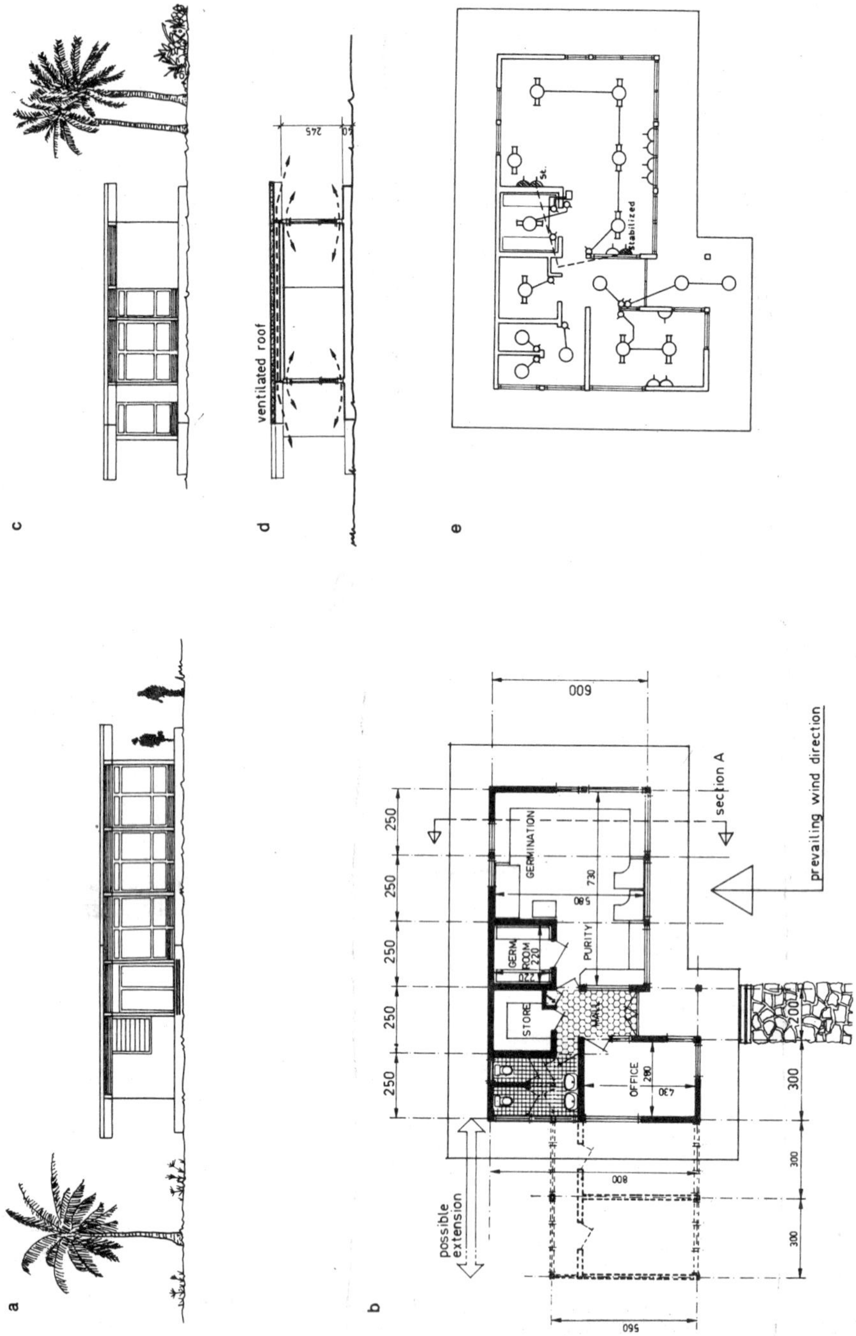

Figure 4. Seedlab 2000, elevations and floor plan.
a. Front elevation; b. floor plan showing possibility of extension; c. right elevation; d. section A of 4.b showing ventilation; e. electrical wiring scheme showing stabilised power supply. Dimensions in cm.

4.3. Ventilation

Roof. The laboratory has an overhanging roof to provide shadow.

- Seedlab 5000: The roof has an inclination of 15° and has a ventilation opening running lengthwise at the top. Just underneath the roof a ventilation opening is also provided, creating a chimney-effect that will ensure ventilation even in the absence of a breeze. The cooling effect of this ventilation system can be increased considerably by fixing aluminium foil at a distance of 3–8 cm below the roof-plating, leaving openings at the bottom and the top to allow ventilation. This design avoids any heat radiation from the roofing material to the air and building underneath. Ventilation openings are provided in the ceilings of rooms, so that air can move upwards to the top of the roof.

  The ventilation opening at the top and just underneath the roof should be covered with 4 mm to 8 mm square wire mesh.

  This type of roof is particularly suited to humid regions, whereas in dry climates a concrete roof, as used on Seedlab 2000 is satisfactory. In view of the size of Seedlab 5000 thermal effects on a concrete roof for this building should be taken into account and attention to detail in construction is essential.
- Seedlab 2000: This structure has a concrete roof, which should not pose any problems in view of the small dimensions of the building. This roof is also ventilated, the space directly underneath the roof having slits between the beams.

  With this construction it is very advantageous to add securely fixed plates of aluminium or another material, some 5–7 cm above the concrete, to ensure free air movement between the layers. This measure reduces thermal effects from the concrete roof considerably.

Walls. For the window-sections a standard size has been chosen in order to enable standardisation of most parts such as window panes, ventilation grids, etc. Prefabrication of these sections incorporating two ventilation openings in the window pane can be advantageous. Each section is composed of several modular parts, from bottom to top:

A – 30 cm, can be fitted with a variable or fixed ventilation opening, or be solid;
B – 60 cm, usually solid;
C – 120 cm, window, can be a louvre-type or two 60 cm opening frames (for the purity section it is advisable to keep windows closed);
D – 30 cm, can be fitted with a ventilation opening (variable or fixed) or an air-conditioned unit;
E – 5 cm, facia board.

The various ventilation openings should all be covered with mesh or mosquito mesh to stop the intrusion of insects, mice, etc. The ventilation openings at the bottom will ensure an effective cross-ventilation, usually considered very effective.

In the central office openings at the bottom of the walls and openings in the ceiling provide ventilation. In addition ceiling ventilators, working upwards, may be installed in the openings where considered desirable.

It is advisable to ventilate the toilets directly to the outside.

4.4. Special installations

For electric equipment such as balances, blowers and the germination equipment a voltage stabilised power supply may be required to ensure a stable voltage (figures 1a and 4e). Sockets should be located for easy use and it is preferable for those for stabilised power to have a different colour or design than those for non stabilised electric current.

In the purity section a weighing table with stone slab or a concrete working bench is required to provide a stable base for installing the balances. This can be purchased or be made on site; special attention should be paid to ensure a vibration-free construction (figure 11).

When handling treated seed an exhaust system must be used, especially when subdividing (submitted sample) and analysing a working sample for purity; also when counting and planting the seed for germination. For instance, a hood could be installed above the place where the seed is handled through which the dust and fumes are sucked off by means of a ventilator.

For the construction of the germination rooms see chapter 9.

Reinforced concrete columns and brick or concrete block walls are planned. If this proves to be too expensive, wood may be used. In that case the roof of Seedlab 2000 also has to be made of wood. All wood should be treated to improve its resistance to moisture, rot and insects.

Construction may be adapted to local circumstances, but it is imperative that the lay-out of the building remains unaltered, in order not to affect the efficiency of the work.

5. Sample reception

The main tasks of the office for sample reception are to register the submitted samples, provide them with an analysis number (= identification of the sample) and to decide what kinds of tests are required. Working sheets (analysis forms) are prepared for each test.

Most samples arrive at the laboratory by post. Preferably, applicants put their requirements on standard forms provided by the laboratory ('request forms', see paragraph 13.2).

Before starting to unpack a sample, it should be verified that both the sample and the tests requested meet with the conditions (such as those in the ISTA Rules) regarding identification, marking, sealing, packing and weight. Irregularities may be found, such as:

- the species or cultivar name on the request form is not the same as on the label of the sample;
- a moisture test has been requested, but a special moisture test sample has not been submitted;

– too much time appears to have elapsed since the sample was taken from the lot.

Such irregularities are recorded on the request form. The applicant or the sampler can be asked to provide additional information. All relevant details are recorded on the form. Preprinted forms of various types are an invaluable aid. Examples of such forms are given in chapter 13. Any information recorded on such forms is readily available to the laboratory.

The applicant's identity should never be shown on the analysis forms as it may bias the performance of an analyst.

Date of receipt and sample registration number are stamped on the analysis forms, the request form and the sample label. The use of a numbering machine and rubber stamps is recommended not only for numbering and for dating, but also for other items that appear frequently, such as codes and species names.

Blank forms, rubber stamps, scissors, glue, etc. should be stored readily available for use in the sample reception section. Stocks should be maintained at a level which will last a season. There should be ample room for temporary storage of incoming samples to leave working surfaces free for the safe handling of each sample and the preparation of the appropriate documents.

6. Administration centre (office)

Each sample, together with the set of forms that has been prepared in the sample reception area is passed to the administration centre. The superintendent checks the partly completed forms and distributes the samples and the forms to the appropriate sections that are to provide the required data. The cover form (see paragraph 13.3.), with the request form, is filed in an open tray, to be readily at hand for consultation. As soon as the other forms have been completed with the test results, they are filed in the cover form. This procedure is necessary in order to be able to provide interested parties (e.g. the sender of the sample) at all times with data on the progress of the different tests, and with provisional test results.

Calculation of the test results may either be carried out by the administration centre or by the leading analysts of the respective sections.

The purity test results are transformed into weight percentages with the aid of a calculating machine. The duplicate percentages are averaged and checked against ISTA tolerances. If they are not in tolerance, additional duplicate tests are prepared.

The germination test results are expressed in percentages based on number of seeds. If the test results are out of tolerance, the germination section performs a repeat test requiring further pure seed and another form.

As soon as all data have been obtained, the results can be forwarded to the applicant. Before that, a final check is made on the completeness and correctness of the data. Then the data are typed on a standard report form or certificate. Different types of standard forms are kept in stock in the office. When the results have been issued, the forms are filed in the cover form and put into the filing cabinet under 'concluded tests'. They are kept for a period (which may depend on legislation) after which they are disposed of.

The administration centre not only ensures the smooth running of the station, it also renders the best possible service to the applicants. This means that (preliminary) results are transmitted to the applicant whenever requested by telephone or by any other means.

7. Moisture unit

7.1. Introduction

The object of a moisture analysis is to determine the moisture content of a seed lot at the time it was sampled. To that end, the sample must be handled in such a way that its initial moisture content is retained. Packed in a sealed moisture proof container (metal or plastic), it must be submitted to the station without delay and analysed directly upon arrival (see also paragraph 13.4.).

Methods for determining moisture in seed are:

A. Oven method. This is the common standard method. For detailed specifications see ISTA Rules. The principle of this method is the elimination of water from the seed by heat under precisely controlled conditions.

B. Quick methods. A variety of brands and types of rapid moisture meters are available. The quick test method should be calibrated or checked against the standard oven method, and is in general less accurate than the oven method.

7.2. Oven method

The moisture analysis is carried out on independently drawn duplicate working samples, weighed with an accuracy of 1 mg. With the exception of cereals (two hours) and maize (four hours) most species are dried for one hour at 130 °C. Seeds containing oil are dried for 17 hours at 103 °C. Each empty container (non corrosive metal) is weighed, including its lid. The submitted sample is thoroughly mixed with a small spoon and two 5000 g portions are weighed directly into the containers. After weighing the containers with seed are placed in the oven which has already been heated to the drying temperature. The temperature in the oven drops when the samples are placed in it and therefore the drying period is counted from the moment the oven has regained the required temperature. At the end of the drying period the lids are placed on the containers and the containers are allowed to cool for 30–45 minutes in a desiccator (filled with e.g. silicagel) and then weighed again. All weighings should occur with an accuracy of three decimal places. The moisture content (M) is calculated to one decimal place by means of the formula:

$$M = \frac{M2\text{-}M3}{M2\text{-}M1} \times 100 = \frac{\text{loss of weight}}{\text{initial weight of seed}} \times 100$$

M1 = weight of empty container with lid

M2 = weight of container with lid and seed before drying

M3 = weight of container with lid and seed after drying and cooling

Where loss of weight can also be calculated as: weight of seed before drying – weight of seed after drying.

Figure 5. Moisture apparatus.
From left to right: moisture proof packed samples; grinding mill; desiccator; aluminium sample containers; pair of tongs; oven with thermometer inserted in top.

The duplicate result may not differ by more than 0.2% moisture content. Otherwise the analysis must be repeated.

Grinding/pre-drying. Some seed species must be ground (see ISTA Rules) before the actual moisture content is determined (cotton, rice, maize, cereals, sorghum, peas and beans). For cereal and cotton seeds fine grinding is necessary; at least 50% of the ground material shall pass through a wire sieve with meshes of 0.5 mm and not more than 10% shall remain on a wire sieve with meshes of 1.0 mm. For leguminous seeds coarse grinding is necessary; at least 50% of the ground material shall pass through a wire sieve with meshes of 2.0 mm. Adjust the grinding mill to obtain particles of the required dimensions. Grind a quantity of seed greater than that required for the test (about 20 g). For the seeds that have to be ground prior to drying, a retest should be made when the moisture content proves to be 17.0% or higher (for soya bean and rice these percentages are 10.0 and 13.0 respectively). But then the seed must be pre-dried before being ground. To that end two 50.00 g portions are weighed and placed on two open trays in the oven at 130 °C for 5–10 minutes (in the case of very moist seed, above 25.0% moisture content, the seed is spread in two open trays and dried at 70 °C for 2–5 hours, depending on the initial water content). The open trays are then placed for at least two hours in the laboratory and each of the duplicate quantities is weighed and a portion ground (e.g. 20 g). The ground material is then subjected to a moisture test using the oven method. The moisture content (M) in the case of pre-drying is calculated according to the formula:

$$M = S_1 + S_2 - \frac{S_1 \times S_2}{100}$$

S_1 = percentage of moisture lost by pre-drying (stage 1)
S_2 = percentage of moisture lost by the oven method (stage 2)

Moisture containers. Containers should be of non corrosive metal (thickness approximately 0.5 mm) with side rounded at the base and a flat bottom, with loose fitting lids that are at the same time very flat in order to prevent loss of moisture. Suggested dimensions are: height 3 cm, diameter base 6 cm, so that there is not more than 0.3 g of material per cm^2. To ensure fitting of the lids, the rim of the container should be levelled by rubbing with an abrasive. Both lid and container should be provided with the same number to identify each sample after the drying period.

Oven. An electrically heated oven with adequate ventilation and thermostatic control which permits the temperature to be maintained at 130° ± 3°C. The heating capacity of the oven must be such that after preheating to a temperature of 130°C followed by opening and loading with containers, the oven will reach 130°C again within approximately 15 minutes.

Balance. A balance which weighs accurately in grams to three decimal places. Although the analytical balance of the weighing unit can be used, a more suitable type is the 'precision balance 2' (the intermediate type of table 1, see paragraph 8.2.).

Grinding mill. This must meet the following requirements:

a. It should be constructed of a material that cannot absorb moisture. Wood is not suitable.
b. It should be constructed in such a way that both seeds to be ground and the resulting ground material are protected as much as possible from the atmosphere of the room while grinding.
c. It should grind evenly, and should not be operated at such a high speed that the ground material is heated. Air currents that might cause loss of moisture must be reduced to a minimum.
d. It should be suited to large, small as well as hard seeds and the fineness of grinding should be adjustable; thorough cleaning should be easy.

Sieves. A set of three wire sieves and bottom receptacle should be available. The set shall have sieves of 0.5, 1.0 and 2.0 mm wire mesh.

Desiccator. The desiccator should have a thick metal or porcelain plate to speed cooling of the containers and seed. The bottom compartment has to be filled with a suitable desiccant, e.g. silicagel coloured with cobalt chloride as indicator: as soon as the dark blue colour becomes pale pink, the desiccant should be reactivated by heating in the oven (up to 130 °C).

A pair of crucible tongs. To handle the (hot) containers.

7.3. Quick methods

Various equipment has been designed to shorten the time taken for moisture determination. These quick methods may be classified by the two different principles used:

7.3.1. Apparatus where the seed is heated directly by an infrared lamp and weighed by a built-in balance. When seeds are heated by an infra-red lamp, heating of the material to higher temperatures is required than for the oven method. Most of these devices have a balance which continuously measures the loss of weight of the sample while being heated. The moisture percentage is usually read off from a direct reading scale. No calculations have to be made. The test can be finished in 10–15 minutes, depending on the kind of seed.

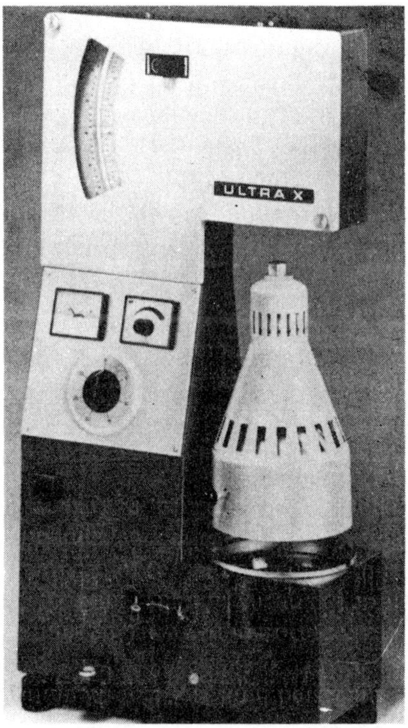

Figure 6. Quick moisture meter with infrared lamp.

Figure 7. Portable, battery operated quick moisture meter.

7.3.2. Electric moisture meters, where the moisture of the seed is directly determined by e.g. its conductivity. Electric moisture meters are frequently used for a quick test. These meters do have a great advantage over all other methods in speed of use. The test can be completed in one minute.

For both types of quick meter each meter must be calibrated for each species and possibly each cultivar. The calibration should be carried out against the standard air oven method and the moisture determinations should be executed under standardised conditions. In general the reading of the meters is less precise than the result determined with the oven method and there may be as much as 1–2% difference.

Consequently, even calibrated meters are only suitable for approximate determination of the moisture content. The meters are especially suited for measuring moisture content in the field and during the process of seed drying and storage in the warehouse.

8. Purity section

The purity section makes tests on the seeds as received. The various functions are located in four units: subsampling, weighing, blowing and purity analysis.

8.1. Subsampling unit
Subsamples are taken from the submitted sample for the purity analysis. The object is to obtain a subsample (working sample) that is of the same composition as the submitted sample but will take less time to analyse.

The ISTA Rules contain detailed prescriptions for subsampling equipment and procedures. For taking a purity working sample (preferably two half working samples, independently drawn), the Wageningen station uses a combination of the mechanical dividing method and the spoon method.

This combination suits the size and the composition of almost any submitted sample, and moreover, enables the operator to work with optimal speed and efficiency.

The object of using a mechanical divider is to minimise any bias in the subsampling. Thorough mixing of the sample can be achieved by passing the entire sample through the divider several times before dividing off the portion required.

The object of further reducing the subsample, by means of the spoon method, is to get as close as possible to the prescribed minimum weight of the (half) working sample.

Various types of mechanical dividers are given in the ISTA Rules. The soil divider is recommended, consisting of the divider as such, and three pouring pans: A, B and C (figure 8).

In using the soil divider, the operator performs as follows:
- he empties the submitted sample bag into the pouring pan A;
- scatters the seed evenly in the pan;
- places the two receiving pans B and C alongside the soil divider;
- empties the pan A into the hopper, allowing the seed to flow at about equal rates along the entire length of the hopper, and to fill B and C each with one half of the submitted sample;
- replaces the pan B by the empty pan A;
- empties the pan B into the hopper, in the same way as previously with A, making equal parts (each 1/4) flow into C and A;
- replaces plan A by the empty pan B;
- empties pan A into the hopper, etc.

In that way a submitted sample is subjected to a series of halvings. The operator continues until a subsample which is five to ten times the prescribed weight of a (half)

Figure 8. Soil divider.
The first of a series of halvings is shown.

working sample is obtained. Then the operator changes over to the spoon method, using the tools shown in figure 9.

The operator:

- pours the subsample evenly on a shallow tray, making it an even layer by pouring carefully with a side-to-side swing; shaking the tray is not permitted;
- with a spoon in one hand, a straight-edged spatula in the other, and using both, small portions are taken and transferred to the weighing beaker on a balance, carefully observing the increase of weight, and taking care that each spoonful is not more than one-tenth of the prescribed working sample weight, and that the spoonfuls are taken from all over the subsample. The latter precaution is to prevent a bias caused by possible segregation of the layer of seed. Segregation may occur both in horizontal and in vertical direction. Therefore, spoonfuls should be taken from all over the tray, and the spoon should be made to scrape over the bottom of the tray, i.e. the layer should not be skimmed, and the spoons should have a straight edge (figure 9);
- stops filling the weighing beaker as soon as the prescribed weight has been reached. The prescribed weight is a minimum and 5% over but not under the weight is acceptable. If it is over, the operator is not allowed to remove the surplus from the weighing beaker. It should either be left, even though the sample will take a longer time to analyse, or the process should be started over again. In the latter case, after emptying the weighing beaker back into the submitted sample bag, it can be refilled, with about 10 spoonfuls, from the material left in the shallow tray;
- writes the weight on a purity analysis form. This will serve as a check, when the working sample has been subjected to a purity analysis and the component weights have been added up. The initial weight and the sum of component weights may differ, the cause of which may not be obvious. If the difference is greater than a certain figure, a further test should be made. A suggested limit is 2% of the initial weight;
- transfers the working sample from the weighing beaker into a purity working sample container, and takes it to either the blowing unit or the purity unit, as required;
- returns the remainder of the submitted sample into the bag, either (a) to be stored in the case of a whole working sample, or (b) to provide for drawing independently the duplicate half working sample which may be needed to confirm the first test, when two duplicate half working samples are analysed.

8.2. Weighing unit

Balances intended for weighing samples, subsamples, fractions and components must meet certain requirements regarding precision. These are specified in the ISTA Rules, as follows:

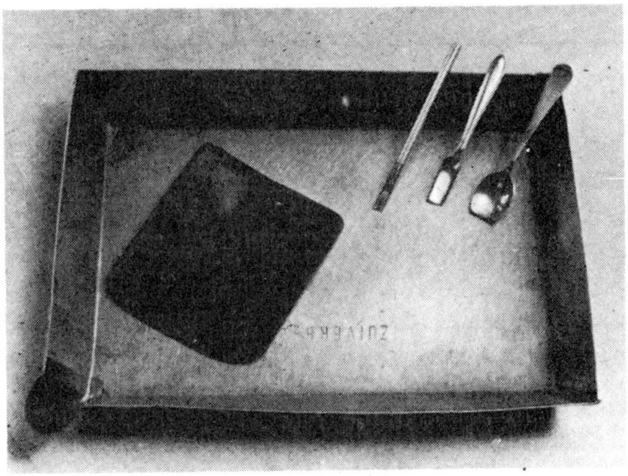

Figure 9. Tools for the spoon method.
Tray; straight-edged spatula; three straight-edged spoons.

When the number of grams specified for the working sample is:	...weigh the working sample and its components to the following number of decimal places:	Examples (g)
less than 1	4	0.8036
1 to 9.999	3	8.036
10 to 99.99	2	80.36
100 to 999.9	1	803.6
1000 or more	0	8036

The whole range of weights between 0.5 and 1000 g can be determined with two balances: an analytical balance enabling 0.1 mg to be read accurately (capacity 160–200 g), and a precision balance with about 1 kg capacity, allowing 10 mg to be read accurately ('precision balance 1' in table 1).

If only two balances are available, congestion of work may occur in the peak season. Whenever congestion becomes a serious problem or when many samples have to be tested for moisture, a third balance should be available; this should be of a type intermediate to the other two ('precision balance 2'). All three balances should be of the direct reading type and provided with a tare mechanism.

Balances (figure 10). The choice of make of balances should mainly be based on the after-sales-service which can be offered by the manufacturer. A firm should be chosen which is able to offer a sound service contract. Servicing should preferably occur twice but at least once a year.

Table 1

	Capacity (g)		Readability (mg)
Analytical balance	160	(200)	0.1
Precision balance 1	1000	(2000)	10
Precision balance 2 ('intermediate' type)	160	(220)	1

Weighing table (figure 11). All balances should be placed on a weighing table. Such a table should consist of a stone slab (8 cm thick) resting on anti-vibration cushioning and supported by concrete or brick pillars. Weighing tables can be purchased from the balance manufacturers, but will usually have to be made locally. A concrete floor or a brick wall are needed at the place where the weighing table will be located.

Figure 10. Analytical balance.

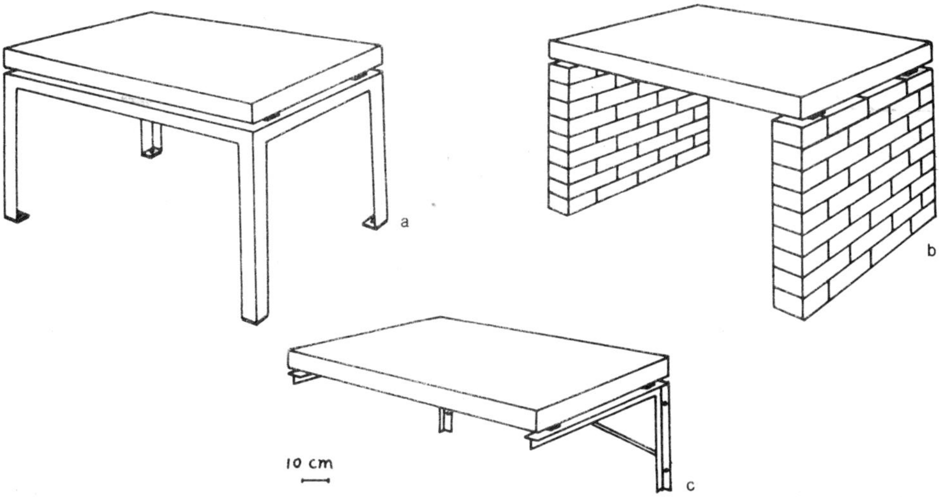

Figure 11. Weighing tables.
a + b. two possible constructions in case the table can be placed on the ground; c. alternative for a situation where no stable floor is present. Note the rubber blocks placed between the stone slab and the base.

8.3. Blowing unit

When grasses have to be tested, then each floret ('seed') has to be checked over a diaphanoscope for the presence or absence of a caryopsis (see purity unit). This work is very time consuming, and the labour costs of such an analysis would be much too high if it were not for the availability of blowers. The blower (figure 12) will separate the working sample into two fractions: a light and a heavy fraction. If the blower is set properly the heavy fraction will only contain florets with a caryopsis; the light fraction then will contain mostly empty florets but also some full ones. Only the light fraction has therefore to be checked and the full florets transferred to the heavy fraction, the remainder being inert matter. In this way at least 50% of the time spent on the analysis is saved.

For certain species ISTA has devised an alternative: the Uniform Blowing Method. To this end the ISTA Secretariat distributes calibration samples (+ instructions). The blower is set with this sample and after blowing the samples to be tested, the heavy fraction is considered to be full and the light fraction is considered to be empty. No checking over the diaphanoscope is needed. This, of course, saves even more time, and is also very benificial for the uniformity in results. The method is compulsory for *Poa pratensis* and *Dactylis glomerata*; it is recommended for *Chloris gayana*. It is the intention that more calibration samples should be prepared for the more difficult tropical range grasses: e.g. *Cenchrus ciliaris, Paspalum* spp., *Panicum maximum*. Blowing apparatus essentially consists of a centrifugal blower, the outlet of which is connected to the bottom end of a vertical tube of a few centimetres internal diameter and about half a metre length. A fine wire gauze, at the bottom end of the tube, retains

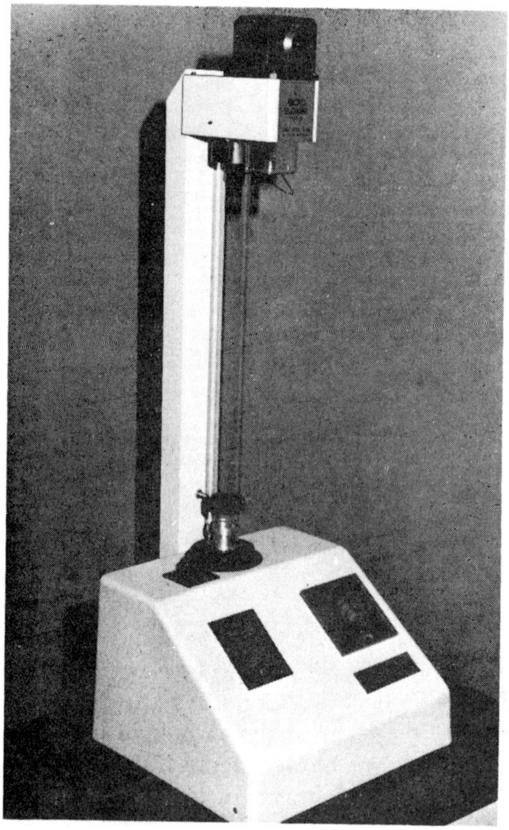

Figure 12. Seed blower.

the sample before it is blown, and also holds the resultant heavy fraction. Different arrangements exist to catch the light fraction. A valve allows the wind velocity to be set at a rate that has been found optimal for the kind of seed.

8.4. Purity unit

The object of the purity analysis is to determine: (a) the composition by weight of the sample being tested and by inference the composition of the seed lot, and (b) the identity of the various species of seed and inert particles constituting the sample. In accomplishing that objective, the sample is separated into three component parts: (1) pure seed, (2) inert matter, (3) other seed. Definitions of these categories can be found in the ISTA Rules. Detailed descriptions of all particles to be regarded as pure seed are given in the Pure Seed Definitions (see also: Handbook on Pure Seed Definitions).

The procedure for purity analysis (apart from subsampling, weighing and blowing), can be described as follows:

- The subsample (purity working sample) is spread on the working table, on either a 'slate' or a 'diaphanoscope' (see below);
- Each particle is judged individually, the criteria used being: external appearance (shape, size, colour, gloss, surface texture) and/or appearance in transmitted light;
- All other seeds and inert matter particles present are removed leaving the pure seed, the separations resulting in the three components mentioned above;
- Each component is weighed (in the weighing unit), the weights being entered on the purity form;
- Components may be retained for future reference although the pure seed will be sent to the germination unit, where 400 seeds of it will be used for the germination test.

An analyst must accomplish separations as prescribed in the ISTA Rules or as prescribed elsewhere. This means judging on sight and moving by hand each particle in the sample. As a purity working sample consists of more than 2000 particles, the amount of time required for carrying out the test is considerable. Therefore it is of great importance that the analyst is well trained and adequate equipment is available to reduce the time taken to a minimum.

Training should enable an analyst to:

1. identify all crop species and all weed species that occur frequently in the samples submitted to the laboratory; any gap in knowledge should be filled by consulting a suitable seed collection kept ready at hand;
2. put the available equipment to the right use;
3. be fully aware of the economic and other implications of test results.

The analyst's equipment should allow working with a minimum of effort and time, and also with a minimum of eye-strain and other tensions or stresses.

A purity working sample containing all three constituents equally and evenly dispersed throughout will take the longest to separate into three components, as every particle has to be isolated and moved individually. Conversely a sample consisting only of pure seed does not need any separation. The more a sample approaches homogeneity, the larger will be the average number of particles that can be shifted simultaneously, and the less will be the effort required. For that reason, whenever feasible, a working sample should be treated so that its components segregate as much as possible, before the actual analysis is commenced. Analysis of the fractions, one after the other, will take less time than the analysis of the sample as a whole. In addition to other suitable methods (see blowing unit) fractionating with a small sieve shaken manually is advantageous.

An optical aid, such as a magnifier or a binocular microscope, should allow samples of seed, spread on a flat surface, to be scanned rapidly. Rapid scanning requires (a) essential details to be definite, and (b) a sizable portion of the sample to be viewed

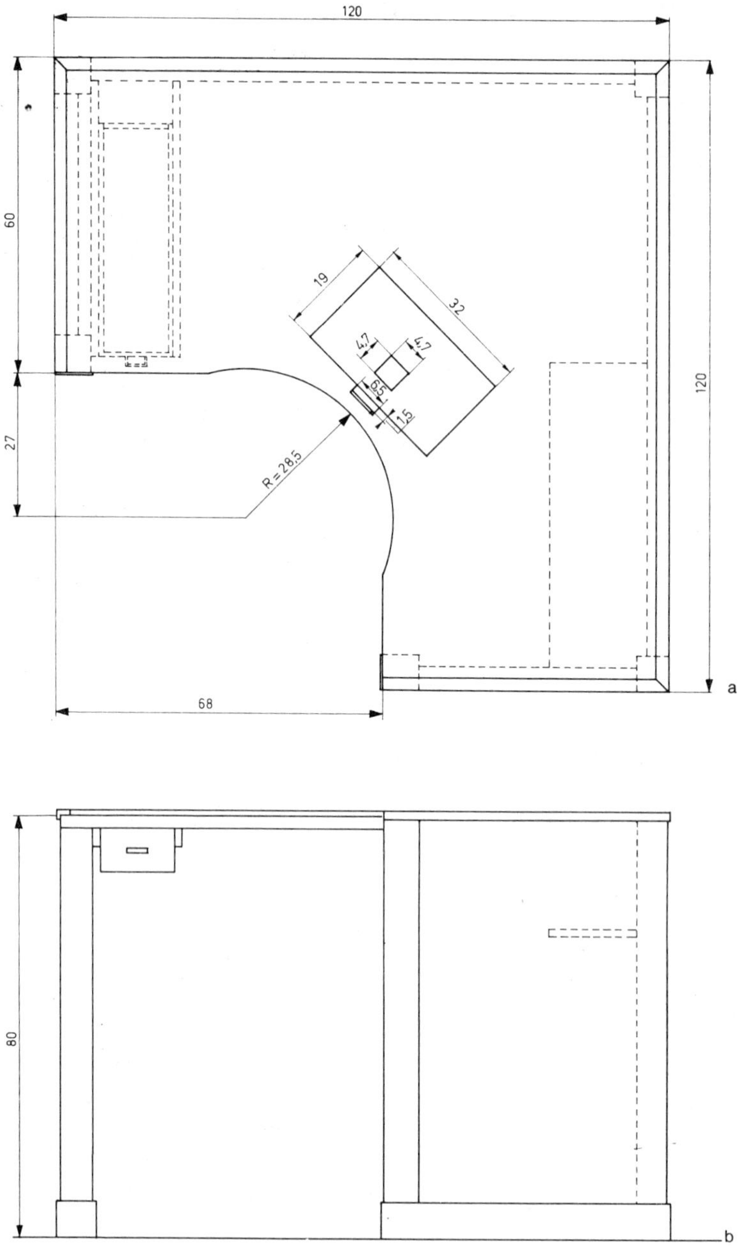

Figure 13. Purity working table.
a. top view showing round cut-out, frame for glass plate, square hole of diaphanoscope and oblong hole of small funnel; b. side view showing small drawer. Dimensions in cm.

at the same time. Experience shows that two magnifications, 3 × and 16 ×, are adequate for virtually all routine purity tests. A suitable mount or stand for the lens is very important; a hand lens (3 × magnification) is put to optimal use only if mounted on a stand, leaving both hands free for the work. The small working distance, the distance between the lens and the object, of many high-power magnifiers limits their usefulness in routine purity work; a lowpower (16 ×) microscope is therefore preferable.

Working tables (figures 13 and 16). The round cut-out in the table gives a large and comfortable working space (good support for the elbows). Each table can be provided with a built-in diaphanoscope, a device that is not commercially available, but enables seed to be judged in transmitted light. A hole of 4.7 × 4.7 cm is cut into the table top and this is covered by a thick glass plate that fits flush into the table top. Underneath a small 20 W (28 V) microscope lamp can be fitted. An additional funnel, fitted into the table is also recommended. Complete freestanding diaphanoscopes are also available.

Incident light. Although the laboratory windows may provide ample daylight, each table should be provided with a strong artificial light. Desk lamps with fluorescent daylight tubes are recommended. On each table a 'slate' (working board) can be laid over the glass plate of the diaphanoscope to judge in incident light, as shown in fig. 16.

A 'slate' may consist of a piece of hardboard 30 × 50 cm, sandwiched between hard plastic, one side dull black, the other side dull olivegreen. The hard plastic surface should be free from static (electric charges) and should be very smooth.

Binocular microscope. A magnification of 16 × is most convenient. The stand should be light-weight, so that the microscope is easy to handle; a stage that is open in front is recommended. In order to achieve this, it will be necessary to cut a section out of the front part of the ring stage commonly provided. Binocular microscopes on swing-arm stands can also be used, but usually this arrangement is found less convenient.

Magnifier. About 3 × magnification, object field diameter at least 5 cm, image perfectly aplanatic and sufficiently achromatic, i.e. without marginal blurring, distortion and discoloration; mounted on a light-weight base easy to handle and allowing unhampered manipulation to the seed sample on the table.

Metal spatula, tweezers and scalpel needle (figure 14). For moving seeds during the test.

Seed collection (figure 15). A cupboard for storing the main seed collection will be needed.

Metal units containing small plastic drawers are recommended. The seeds are put into labelled glass test tubes and stored in the drawers.

Figure 14. Various small instruments.
a. flat spatula for purity test and counting seeds for germination; b. scalpel needle for same; c. scraper for removing and levelling sand in germination containers; d. tweezers, drawn in two positions.

Figure 15. Main seed collection.

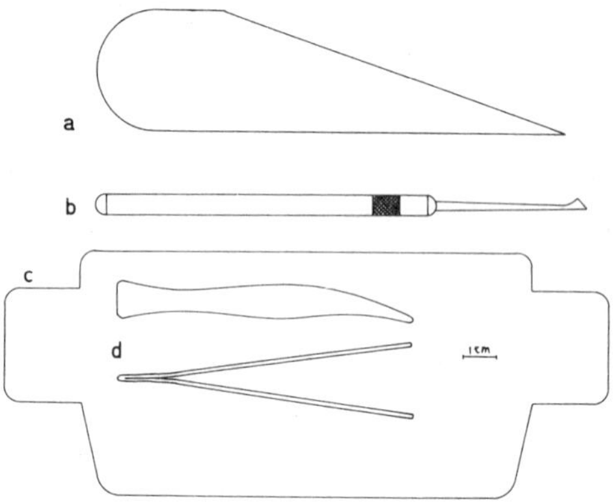

Other instruments to be used (figure 16).

- a 'half funnel', with the aid of which a sample of seed from the table can be directed into a small sample container;
- a set of watch glasses, used to cover seed fractions when the analyst is working on the sample; recommended diameters: 12, 10, 8 and 6 cm;
- aluminium containers; two sizes: 10 and 20 mm × 40 mm diameter respectively. The larger one may contain the pure seed fraction, while the smaller ones are designed for the other fractions. The lid of the containers should be well-fitting but should not be tight. The containers themselves should be 10 and 20 mm high respectively; the lid can be as shallow as 5 mm. The containers can be made of aluminium or another metal: plastic and cardboard are not recommended (see also figure 17);
- a small quick reference collection of crop seeds and weed seeds, in specimen tubes that are arranged in holes in a wooden block;
- the chair used should be especially designed for uninterrupted sitting at the bench;
- shelves or cupboard for temporarily storing purity components; its size depends on the intended use and on the number and size of sample components that are stored simultaneously.

Figure 16. Purity working table and tools. For explanation see text.

Figure 17. Aluminium purity containers.

9. Germination section

In this chapter the equipment needed for germination is reviewed, and the requirements it must comply with are explained. Particulars of the equipment and firms from whom it can be obtained are given in chapter 14.

Because the lay-out of the germination laboratory is treated in chapter 4, in this chapter no further attention is given to this aspect. That applies also to administration of germination which is explained in detail in chapter 13. For details on how to execute the germination test and how to evaluate the seedlings, refer to the ISTA Rules and the Handbook for Seedling Evaluation.

9.1. General requirements for germination

1. The relative humidity of the air surrounding the seeds must be kept near 100%. This level must be maintained not only during the long periods of constant temperature but also during the relatively short times of change from one temperature to another when using alternating temperature cycles.
2. An accurate adjustment of temperature within limits of ± 1 °C is necessary. For alternating temperatures a relatively quick change of temperature must be achieved in less than an hour. Germination cabinets meeting these requirements usually have sufficient capacity for cooling and heating to realize a quick temperature adjustment when the doors are repeatedly opened. With the germination apparatus generally in use a quick change in temperature for alternating temperatures can only be achieved when filterpaper is used as substrate. For germination tests in sand changes in temperature take comparatively rather long. However, tests in sand are as a rule made at constant temperatures.
3. A light source must be provided, using white fluorescent tubes, with a relatively low emission in the far red and a high special emission in the red region. The lamps should be installed in such a way that illumination is as uniform as possible and of an intensity of 750 to 1250 lx. Precautions should be taken that the starter and the chokes producing heat are positioned so that they do not affect the control of the germination temperature and humidity.
4. Gas exchange should be possible so as to keep the composition of the atmosphere surrounding the seed approximately normal (continuous supply of oxygen and removal of such gases as carbon dioxide). However experience has shown that artificial, forced ventilation does dry out the tests.
5. The dimensions of the substrate should be such that maximum use of the equipment is achieved. However, spacing between seeds and between tests should not hamper the germination process or cause undue spread of any seed-borne disease.

The most widely used apparatus in germination testing are the germination room, the germination cabinet and the germination table. Important, especially with regard to the room and cabinet type of germinator, is effective temperature insulation. Also important is that the insulation material should be protected from moisture penetration, making it lose its function.

9.2. Germination rooms (figure 19).

Rooms which are only conditioned for temperature (constant or alternating) by an air-conditioning unit are more and more commonly used now. Maintenance is relatively cheap and simple.

Control of light can be arranged in such a way that only those lights in the part of the room that is in use can be switched on. As mentioned already the rooms should be well temperature insulated. The purpose of insulation used in these controlled rooms is twofold. First, in order to assure the most efficient air-conditioning the inflow of heat from the outside should be reduced to a minimum. Similarly when outside temperatures are lower than the room temperature (at night) loss of heat should be minimised. Secondly, condensation of water on or (worse) inside the walls of the germination room has to be avoided, because this might well damage the wall construction, and cause the insulation material to lose its function.

Construction of walls with a low thermal conduction will solve the first. The second issue poses a more difficult problem. Ideally heat insulation should always be placed on the warm and humid side of a partition. However, this may cause constructional and practical problems and therefore the insulation material will have to be placed inside the germination rooms. Special care in design is then needed, the approach depending on the climate.

Between the insulation material (polystyrene, polyurethane, etc.) and the wall a layer of aluminium foil should always be fixed, in order to minimise heat radiation.

Starting from the presumption that outside temperature is around 35 °C and the temperature inside the germination room needs to be 10 °C, the following construction of the walls is needed (the examples are to be taken as a guideline):

Maximum RH outside (%)	Thickness of insulation (cm)	Thickness of concrete wall (cm)
70	5	15
77	5	10
85	10	10
90	5	open wooden frame
95	10	open wooden frame

In warm and humid climates (over 85% RH and around 35 °C), the optimum solution is to replace the concrete wall by an open wooden frame ensuring sufficient ventilation on the outside (figure 18).

When the ceiling consists of a wooden construction a 5 cm thick layer of insulation material will be sufficient for most cases.

Rooms which will meet all these specifications may be difficult to build locally. In that case it is advisable to build rooms of prefabricated well insulated panels. Nearly all possible sizes are available (modular system). Such panels may be constructed according to the sandwich principle where the insulating material (e.g. polyurethane)

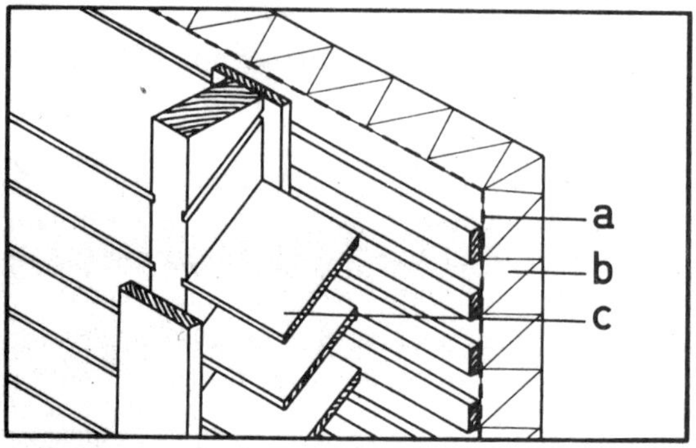

Figure 18. Open wooden frame.
Wall construction of germination room under warm (about 35 °C) and humid (relative humidity over 85%) conditions. a. aluminium foil; b. insulation material; c. open wooden frame. See text.

is foamed between, for instance, enamelled steel plate. The panels can be delivered with varying thickness of the insulation material determined by the likely temperature differences between the room and outside.

Assuming the germination rooms in the 5000 samples laboratory have a maximum capacity of 560 containers for germination tests in sand, it can contain 140 samples of four replicates of 100 seeds (e.g. cereals, rice) or 70 samples of eight replicates of 50 seeds (e.g. maize). For containers see paragraph 9.8. If the 2000 sample laboratory has a germination room which will hold a maximum of 336 containers for germination tests in sand, it can contain 84 samples of cereals or 42 samples of maize. When using paper substrate and the plastic boxes from paragraph. 9.8. (21 × 15 × 3 cm), the rooms of Seedlab 5000 and 2000 will hold 1680 and 1008 boxes respectively. Depending on the species each box can take one or two replicates, meaning that the rooms can accommodate 420 or 840 (Seedlab 5000) or 252 or 504 (Seedlab 2000) samples.

9.3. Germination cabinets (figures 21 and 22).
In the cabinet, as in the rooms, the seed can be germinated in darkness or light. However, light for germination is advised because more sturdy seedlings are produced and in many instances the seedlings are more easily and correctly evaluated. On top of these positive points the red emission of the white fluorescent tubes has a dormancy breaking effect (see paragraph 9.1.3.).

Cabinets may be 'wet' or 'dry' cabinets and be equipped for either constant temperatures or for both constant and alternating temperatures. Contrary to 'wet' cabinets, the tests in 'dry' cabinets must be covered against drying out during the germination period. When using filter paper substrate, the 'wet' cabinet can hold many more samples than the 'dry' cabinet where all tests have to be germinated in containers. Germination cabinets and also germination tables (see below) require more specialised maintenance when compared with germination rooms. With regard to cooling, when the cooling unit is water-cooled, the cabinet needs a continuous supply of clean water. This is also the case when the temperature and humidity conditioning is regulated via water. Consequently, when choosing a cabinet one should take into account

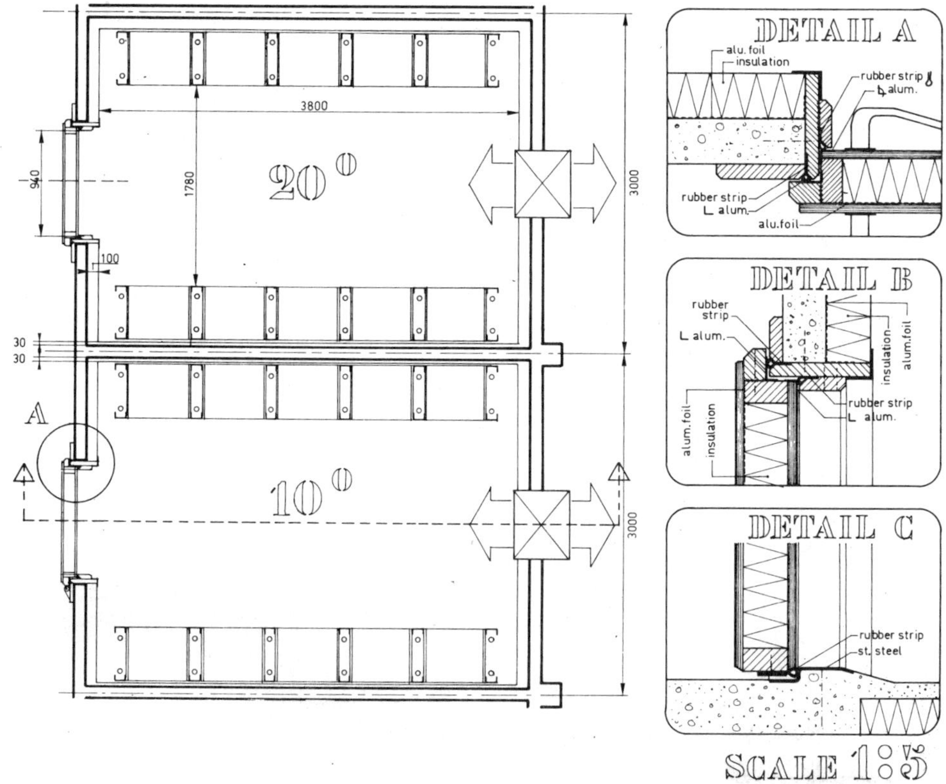

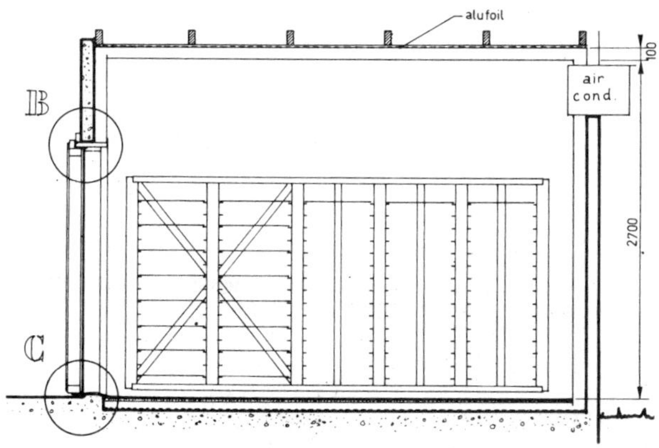

Figure 19. Germination rooms.
top left: floor plan of two germination rooms, showing arrangement of racks, fluorescent tubes, placing of doors and air-conditioners; bottom left: cross section of germination room along broken line as indicated in floor plan above; top right: details of door construction. Dimensions in mm.

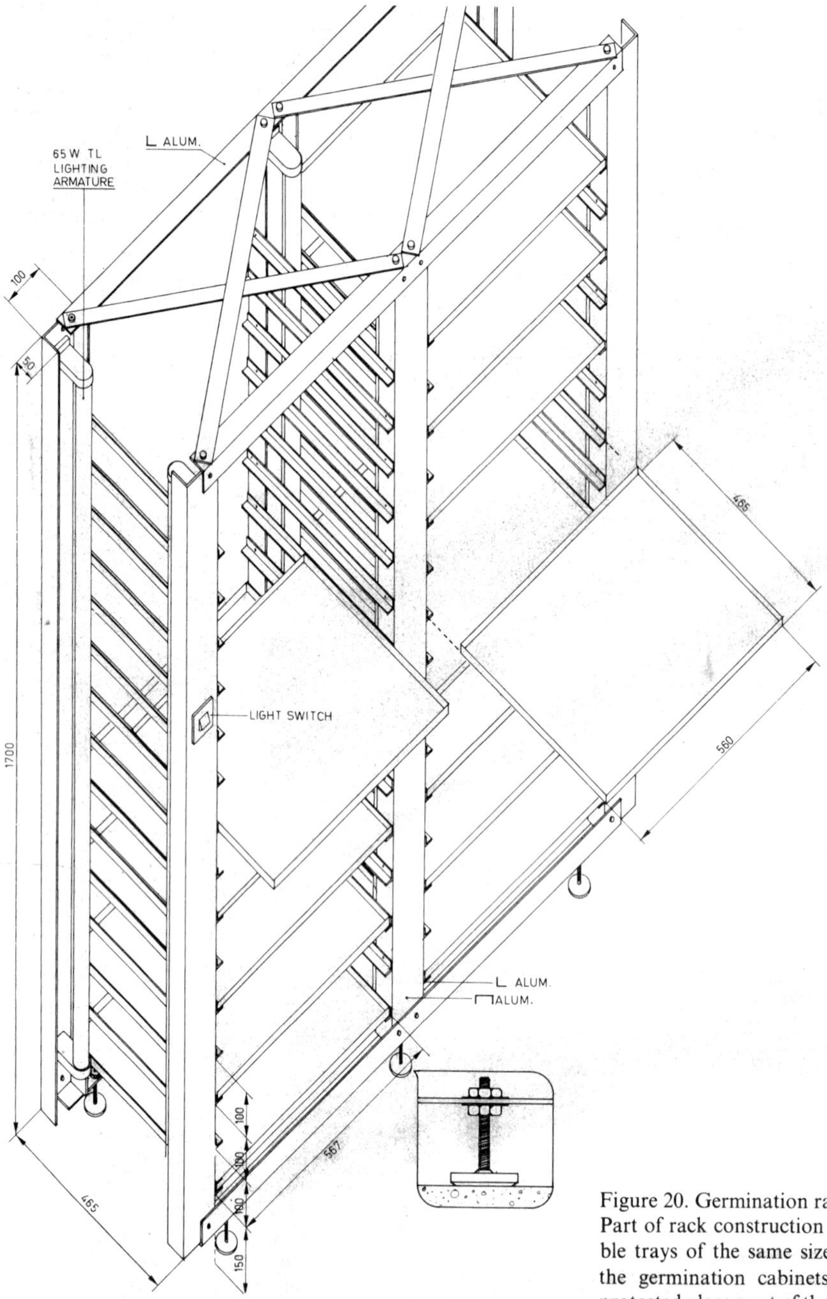

Figure 20. Germination racks.
Part of rack construction with movable trays of the same size as used in the germination cabinets. Note the protected placement of the fluorescent tubes. Dimensions in mm.

Figure 21. 'Wet' germination cabinet with light.

whether a continuous water supply can be guaranteed. Otherwise, the choice should be made for a cabinet which for its cooling and temperature/humidity conditioning is independent of a continuous water supply.

Except the Navep cabinet (see chapter 14) which is only air-cooled, all cabinets advised in this project can be delivered with either air or water cooling. The temperature range of the cabinets, for either constant or alternating temperature conditioning, is between 5° (0°)C and 35°C. They are well insulated and the walls are protected against penetration of moisture into the insulation material. However, in tropical and sub-tropical regions it may be wise to place the cabinets, especially those that are air-cooled, in an air-conditioned room (e.g. germination room) to extend the life of the cooling unit and to bring down the energy costs.

9.4. Germination tables (figure 23 and 24).

The germination table consists of a germination plate upon which filterpaper beds are placed. The seed beds are kept continuously moist by means of a paper wick (figure 24), which extends from the seed bed through slots or holes in the germination plate

Figure 22. 'Dry' germination cabinet with light.

into the underlying water bath. To prevent drying out, the seed bed is covered with a bell jar provided with a hole allowing for ventilation without undue evaporation. The temperature is conditioned directly by conditioning the germination plate or indirectly by heating/cooling the water in the waterbath. The tables advised in this project have direct conditioning, either by the temperature controlled water running in a closed circuit through tubes attached underneath the germination plate or through tubes of stainless steel which make up the germination plate themselves. The water in the water bath may have to be replaced from time to time in case it has become dirty. However, the water used for temperature conditioning need not be replenished because it runs in a closed circuit. When air-cooled, this system makes the apparatus rather independent of a continuous supply of water. The tables may also be delivered water-cooled upon request. The tables are conditioned for constant and alternating temperatures between 5°C and 35°C.

Figure 23. Germination table.

The quantity of germination equipment depends on the following:
a. the number of samples that must be investigated in the peak seasons;
b. the kinds of seed during that time;
c. the germination methods used (e.g. temperature, substrate, germination period, possible pretreatment for breaking dormancy);
d. effective dimensions of the equipment in relation to the space each test replicate takes (beds, containers).

For a 5000 sample laboratory with two germination rooms it may be necessary to have one germination table and at least one cabinet to prechill samples in order to break dormancy. However, when only cereals like wheat are tested, one of the rooms may be used for prechilling. For a 2000 sample laboratory equipped with one germination room, depending on the kind of species tested a cabinet for prechilling may be necessary, and also a cabinet with light or a germination table to extend the possible temperature regimes. Also for research purposes (e.g. methods) cabinets or germination tables are advised as a supplement to the germination rooms.

9.5. Counting devices

For some kinds of seeds counting devices have replaced time-consuming hand counting. In general, smooth, non hairy and round to elliptical seeds can be satisfactorily

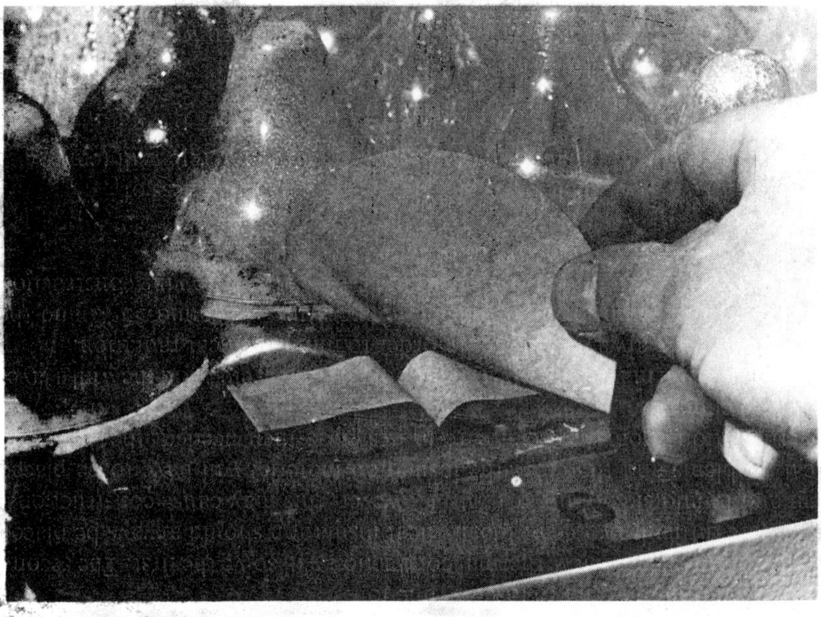

Figure 24. Detail of germination table, showing water supply to the circle by means of a paper wick.

counted with these devices, provided that the seeds are not too small.

There are two basic types:

1. vacuum counting heads, e.g. for seeds of the size of
   a. cereals (e.g. wheat and rice)
   b. brassicas
2. counting boards, often used for counting maize, lupins, peas, beans and, when there is no vacuum-counting head available, also for cereals. Counting boards are relatively easy to make locally.

Requirements for counting devices

a. The size of the counting heads (vacuum counter) or counting boards must correspond with that of the germination substrate. It is important that seeds are spaced regularly and not too close to each other (in order to prevent spread of infection if disease is present on the seed).
b. The seeds are counted in units of 25, 50 or 100, so that the results of the germination tests can be easily computed and checked for their agreement with the appropriate ISTA tolerance table.

In counting segregation of the seeds must be avoided. The way in which the seeds are put on the counting head or board can determine whether segregation will occur or not. Rolling of the seed should be avoided as much as possible. The seed should

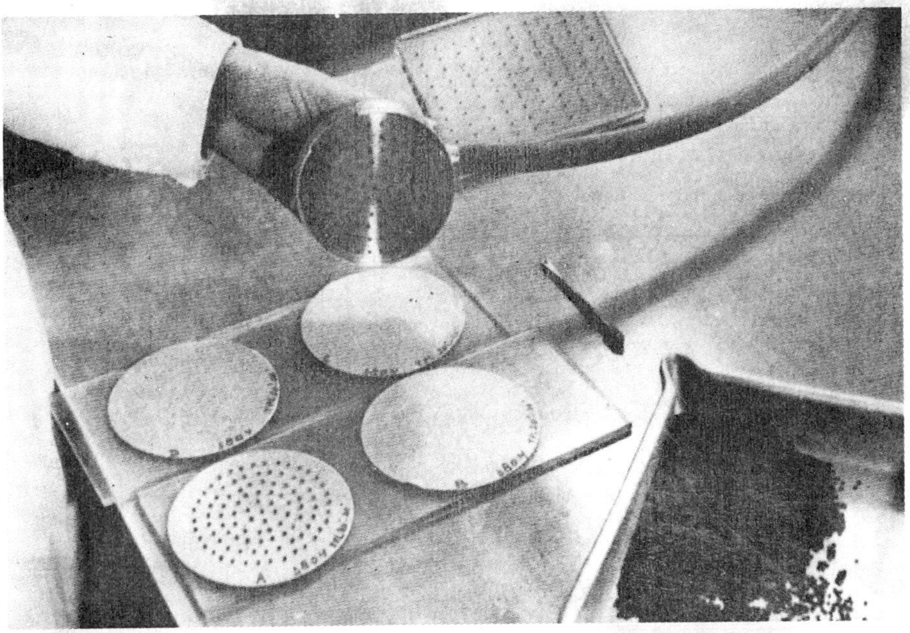

Figure 25. Vacuum counting heads.
Round head for filter paper circles and seeds of the size of brassica. In the background: rectangular head for larger seeds.

therefore be carefully poured out on different spots of the head or board, so that it easily spreads over its whole surface. The vacuum head should never be placed into the seed, as the lighter seeds will be selected. Because counting heads and boards are difficult to clean, it is advisable to use different heads and boards for chemically treated and untreated seeds respectively.

9.5.1. Vacuum counting heads (figure 25). Dimensions of the head depend on the size of the germination boxes or filterpaper, but should always be fractionally smaller. For squares and rectangles: 0.75 cm less in length and width; for circles: 0.75 cm less in diameter.

Diameter of the holes in the counting head varies with the size of the seed e.g.:
1.1 mm for seeds of the size of cereals;
0.3 mm for seeds of the size of brassicas.

Some deviation in hole size is acceptable, because the effect also depends on the strength of the vacuum. The counting head should have an edge to prevent the seeds from rolling off. This edge should be interrupted for some length for removal of the surplus seed. Good results can be obtained by using a high-power vacuum-cleaner.

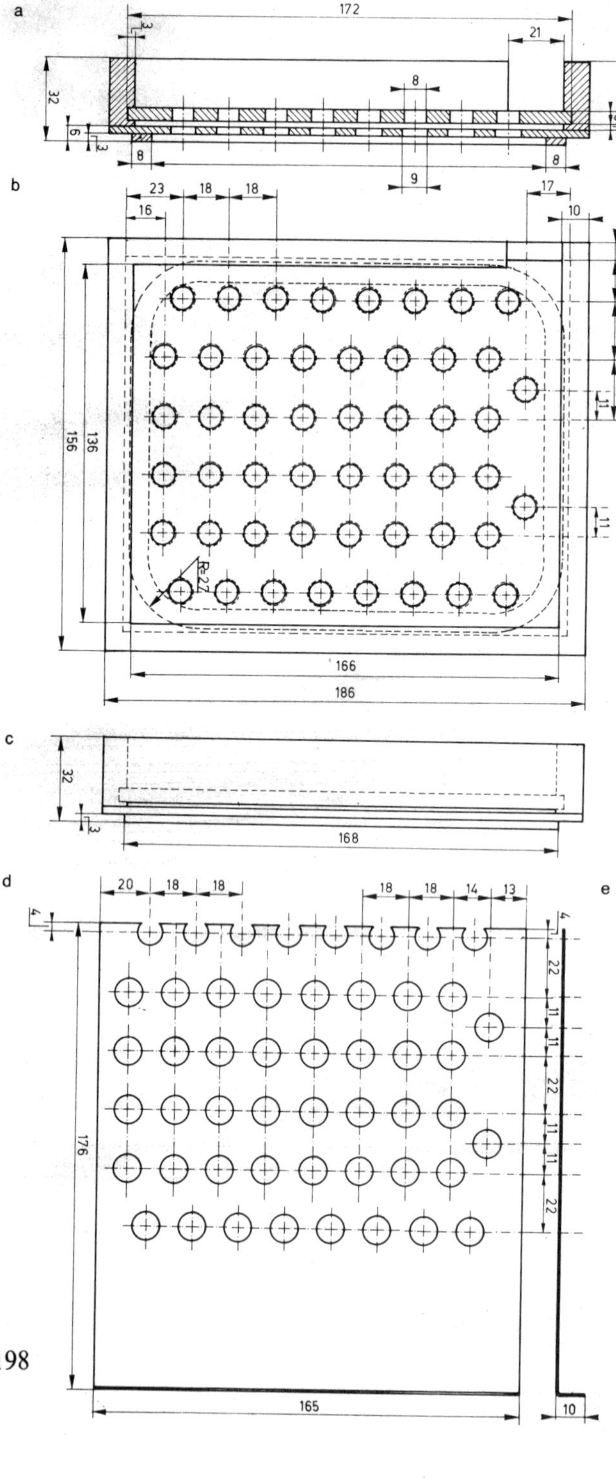

Figure 26. Counting board.
a. cross section; b. top view; c. side view; d. movable sleeve, top view; e. movable sleeve, side view. Dimensions in mm.

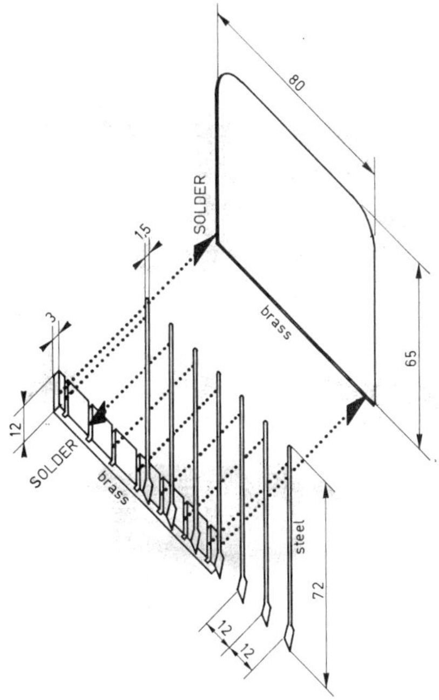

Figure 27. Assembly of germination rake. Dimensions in mm.

Use of vacuum counters:
- Place the counting head horizontally with the holes upwards;
- Close valve and bring seeds on head with the vacuum off;
- Apply vacuum and remove excess of seed;
- Check that there is one seed on each hole;
- Turn counting head over; release vacuum and drop seeds on substrate.

9.5.2. Counting boards (figures 26 and 30). The dimension of the boards depends on the substrate, though the length and width should be 0.75 cm less than the seed bed.

Diameter of the holes varies with the kind of seed (peas, beans, maize, etc.), but must allow the largest seed of a sample to pass through.

The bottom is provided with 25,50 or 100 holes, the number and size of these depending on the size of the seed and the dimensions of the seed bed. In the bottom is a movable sleeve with the same number and diameter of holes. When the sleeve is positioned so that the holes in bottom and sleeve correspond the seeds will fall onto the seed bed.

The counting board should have an edge on all sides, but interrupted at one place for removal of the surplus seed.

9.6. Sand bunker

Sand for germination tests can be stored in a bunker, which shall be made of concrete or bricks. Finishing has to be made smooth: e.g. with cement plaster or tiles. A provision for filling the bunker from outside is very convenient.

9.7. *Kitchen department*

For the cleaning of germination equipment and for preparing drinks for the laboratory staff, household kitchen equipment including a small refrigerator, two sinks, a stainless steel draining board and a water heater is needed.

9.8. *Further needs*

Rakes and scrapers (figures 27 and 14c). Small rakes and scrapers are used for loosening and smoothing the sand beds. At least two of each are required. For cereals and pulses the same type of scraper can be used, but for small seeds such as onions, which are sown more shallowly the scraper must not work as deep.

The method of use in the Government Seed Testing Station is explained below (see also figures 28 to 38).

For cereals or pulses the boxes are filled with sand to such a height that, after using the long side of the scraper to take away the excess sand, a seed bed layer of about 2.5 cm thickness is left. This layer is loosened with the rake and the seeds are sown. They are then covered with sand that is carefully raked, while avoiding touching the seeds. Excess sand is then removed with the short side of the scraper. The resulting structure allows good gas-exchange. For cereals and pulses sowing depth should be 1.0–1.5 cm; for onions 0.5 cm.

Containers. For germination tests in sand, e.g. for wheat, rice, maize and pulses. They measure 14 × 17 × 4.5 cm (figure 39), and can be provided with a tight fitting 9 cm high transparent cover for use under dry conditions (e.g. germination rooms).

For germination tests in pleated paper, e.g. for pelleted seeds and beets. They measure 21 × 15 × 3 cm (figure 40), and are used with paper substrata as recommended in chapter 14. The boxes can be stacked. Separate plastic lids are available.

Bell jars. To prevent drying out the seed beds on a germination table are covered with a bell jar, provided with a small hole permitting ventilation without undue evaporation. A bell jar should not completely cover the seed bed but leave a small edge exposed allowing some evaporation. This ensures transport of possible toxic and colour substances to the edges of the paper substrate.

Filter paper. A general requirement is that the filter paper used in germination tests is free from chemicals that are likely to affect seedling development. In the Dutch Station greyish round paper beds are used; this colour facilitates evaluation of the roots.

Two sizes, dependent on the size of the seed tested, are used, with diameters of 8 and 10 cm. The thickness of these papers is 0.6 mm (code no. T 300).

In the germinator the beds are placed on a layer of filter paper of the same size as the tray. This underlayer is white, and 0.5 mm thick (code no. ZH 1220).

In some instances folded papers of the envelope type are preferable. They are made of white paper folded to 26.5 × 23 cm. The thickness of this paper is about 0.3 mm (code no. ZH 1224). Pleated paper is used for germination of for instance beet seeds (code no. 3014, for the coverstrip no. 0858).

Almost all kinds of seeds referred to may be germinated either on paper (TP) or

Figure 28–38. Planting procedure in sand using a counting board.

Figure 28. Container is filled with moist sand and levelled with the long side of the scraper.

Figure 29. The seed bed is loosened with the rake.

Figure 30. Counting board is evenly covered with seeds.

Figure 31. Surplus of seeds is shaken off.

Figure 32. The number of seeds is corrected with a tweezer.

Figure 33. Counting board is placed over seed bed and metal sleeve is pulled to release the seeds.

Figure 34. Planted seeds are carefully covered with loose moist sand.

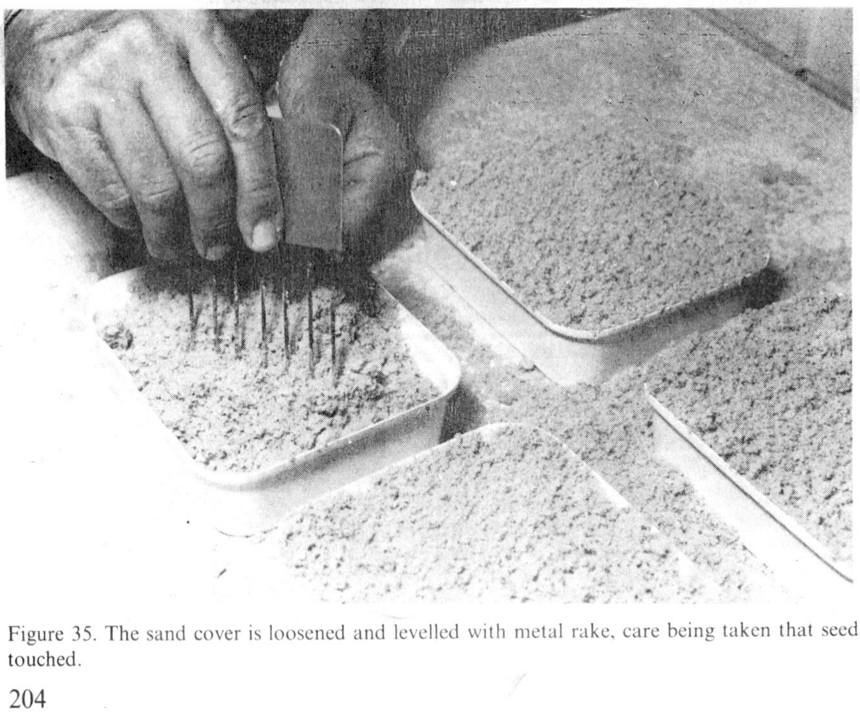

Figure 35. The sand cover is loosened and levelled with metal rake, care being taken that seeds are not touched.

Figure 36. Excess of sand is removed with short side of the scraper.

Figure 37. Slips with analysis number are inserted in beds.

Figure 38. Four of the eight replicates ready to enter the germination apparatus.

Figure 39. One replicate in plastic container with transparent lid, ready for evaluation.

Figure 40. Shallow transparent containers for tests between paper; in this case pleated filter paper.

in sand (S). In the Dutch Station seeds of the size of wheat or larger are germinated in sand but smaller seeds are most often germinated on top of paper. Between paper methods have the disadvantage that the seeds easily make contact, increasing the danger of transfer of disease, and no light reaches the seedlings making them more subject to injury and rotting, and less easy to evaluate.

Whenever local paper is used comparative tests with other officially approved paper or with sand should be performed to check on the suitability of the paper.

Sand. In the Dutch Station white fine sand and river sand is used which does not need to be sterilised. In case the sand available is not free from injurious organisms it should be sterilised. The composition of the sand should be such that it passes a sieve with holes of approximately 1 mm diameter, but should not contain very fine particles like clay dust. After use the sand must either be discarded or it must be sieved, washed and sterilised. Sterilisation of sand can be performed at 150 °C for two hours in open steel pans. The sand should only be re-used a few times because of danger of accumulation of toxic substances.

Watering bottles. In general watering bottles are not used for moistening the germination substrate but can be used to give some extra water to filter pads to facilitate the removal of the seedlings.

Tweezers (figure 14.d). To handle small and tender seedlings short-armed tweezers should be used. To avoid injury the tweezer tips must not be too thin or sharp.

Spatulas (metal or plastic) for counting seeds.

Needles for evaluating seedlings, e.g. the growing point.
Safety razor blades with holder for scarifying or cutting seeds.
Pre-drying oven. This oven is used for breaking dormancy by keeping the dry seeds for a period at 35 °C before planting. The same oven can be used as needed for moisture testing.
Mixer. For mixing sand with water so that it contains the right amount for germination, the use of a small concrete mixer is recommended.
Washing apparatus for beet seed. Washing may be carried out in a special apparatus as explained below or just simply in cloth or nylon bags in running water as long as the requirements mentioned under 1 and 2 below are fulfilled.

The purpose of this equipment is to remove germination inhibitors from beet clusters. It must meet the following requirements:

1. the washing water must be regularly drawn off and replenished;
2. its temperature should be 25 °C constant;
3. the material must not be toxic nor corrosive.

Binocular microscope. The germination unit needs one binocular microscope with the same specification as for purity tests for closer examination of ungerminated seeds at the end of the test and for tetrazolium tests (see paragraph 8.4.).
Chairs. The chairs should be of an adjustable type to prevent tiredness as much as possible.

10. Viability unit (Tetrazolium)

The topographical tetrazolium method is a biochemical test to determine quickly the viability of a seed sample, but can also be applied to individual seeds that remain hard or fresh and ungerminated at the end of a germination test. The staining procedure should be executed in darkness at 30 °C. A small oven may be needed for this reason. Generally, depending on the species to be investigated, a 0.5% or 1.0% solution of 2, 3, 5-triphenyl-2H-tetrazolium chloride ('red tetrazolium') is used, obtained by dissolving 5 g or 10 g of the tetrazolium salt in 1000 ml of water with a pH ranging between 6.5 and 7. If the pH of the water is not in the neutral range the tetrazolium salt should be dissolved in a phosphate buffer solution. The tetrazolium solution must be stored in darkness.

Detailed information can be obtained from the ISTA Rules and the Tetrazolium Testing Handbook.

Remark: The results of tetrazolium and germination tests are generally in close agreement, but although properly conducted large discrepancies are possible. In tetrazolium testing only the embryo is evaluated, disregarding the influence of the outer structures of the seed (e.g. seed coat) which may influence the germination result (disease, dormancy). Besides, contrary to evaluation of the seedling in the germination test, not all abnormalities can be detected in the small embryo. Consequently, unless the seed is of a high quality, the tetrazolium test tends to give a higher result.

11. Storage room

As samples may need to be retested at some later date, for example in case of a dispute, it is necessary to have the availability of a storage room to keep the samples for about one year without losing quality.

To maintain germination capacity the seed must be stored at a temperature of about 18 °C and a relative humidity not higher than 65%. To meet these minimum conditions it may be necessary to use an air-conditioner. Precautions against damage by insects and rodents are imperative especially in tropical regions. The storage room must be made rodentproof and the seed stored in insectproof containers. Samples however that are already badly infected on arrival should be destroyed after testing because it is hardly possible to maintain the original quality during storage and because the risk of infection of other samples is too great.

12. Library

A convenient location should be chosen where books and scientific journals are placed with appropriate accommodation for consultation (see list of suggested books and journals, chapter 15).

13. Handling of analysis results

13.1. Introduction

In this chapter a system based on that used by the Government Seed Testing Station is explained. It consists of six forms: a cover form and five loose forms, all made from stiff paper.

The forms are numbered 1 to 6, and each has its special function.

Form 1: the folded cover form keeps all forms together, and is used for the calculation of the costs of the analyses.
Form 2: the request form used by the applicant for requesting analysis which is sent to the station together with the submitted sample.
Form 3: the moisture form used internally for the moisture analysis.
Form 4: the germination form used internally for the germination analysis.
Form 5: the purity form used internally for the purity analysis.
Form 6: the extra form used for all additional determinations (other seed by number, 1000 seeds weight, etc.).

In the upper right corner of all six forms space is provided for the analysis number. Upon arrival the sample is given a registration number in sequence of receipt. All forms will be given the same analysis or registration number. When the 5 forms are filed in the cover form in the right sequence 1 to 6, the six numbers can be seen at a glance. In this way it can easily be seen whether the 6 forms agree. If any of the

request number	REQUEST FOR SAMPLE ANALYSIS						• *analysis number* 2
copy of cert. to:	Information to be provided by applicant: certificates required	nation.	BIC	OIC	GIC	language	
	number of certificates						
	mark required analysis →	purity	germ.	moist.			rush anal
signature of applicant	name and address of applicant						
	species, cultivar, generation, etc.						
copy of cert. to:	Information to be stated by (official) sampler: sampling and sealing agency						
	official marks of lot						
	official seal of lot						
	weight of lot						
signature of sampler	number of bags or other containers	date of sampling	date of sending sample	•date sample received			
N.B. applicant should keep a copy of the request form and refer to the request number when seeking information from the seed testing station			• to be completed by seed testing station				

Figure 41. Form 2, Request form.
Used by applicant when submitting a sample for testing. For explanation see paragraph 13.2.

analyses are not required one or more of the forms can be omitted.

The forms are depicted in figures 41 to 46. The *italics* used at various places on these forms are not to be printed. The six forms are used as follows:

13.2. The request form (Form 2, figure 41)

This form should be used by the applicant or by the official sampler and should accompany the sample. Regular customers should be encouraged by the station to hold a stock of request forms. These forms should be provided with a pre-printed sequential number in the left corner. This number is not identical to the analysis number put in the right corner of the form. The number in the left corner helps the applicant to describe the sample when inquiring after preliminary results of tests. To this end the seed testing station keeps a record of all request numbers with their respective analysis numbers. The sender of the sample should retain a copy of the form. It is divided into three sections. The first (above the double line) contains the information provided by the applicant. This and the information stated by the official sampler (below the double line) provide the details for completion of ISTA certificates (Orange and Green International Seed Lot Certificates, abbreviated: O.I.C. and G.I.C.; the

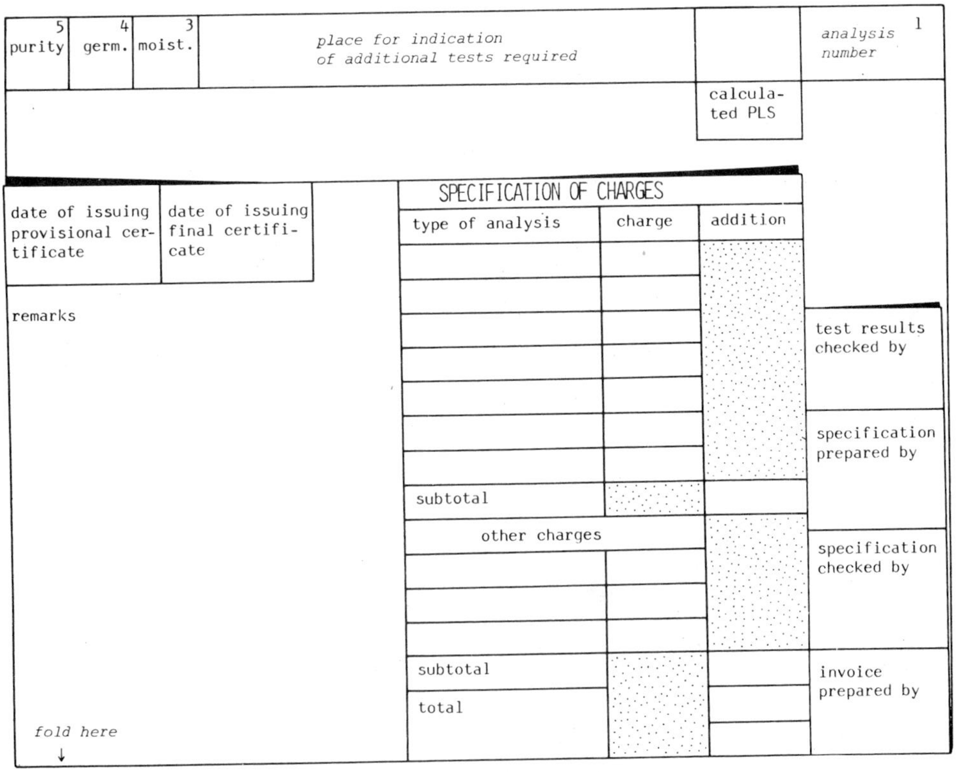

Figure 42. Form 1, Cover form.
For filing the analysis forms and request form, and used for calculating the charges. For explanation see paragraph 13.3.

third one being the Blue International Seed Sample Certificate, or B.I.C.). The third area (right of the thick line) is for use by the Seed Testing Station.

The applicant can tick the kinds of analysis wanted (purity, germination, moisture, etc.) and can ask for the analysis results:
- on national or international certificates;
- in specified languages;
- within a certain time (rush analysis).

Seed legislation of many countries requires prescribed tests.

Prescriptions for using ISTA certificates require the name and address of the applicant, some data about the seed lot such as species, cultivar name and weight of the lot. It is advisable that there should be provision for the applicant to sign the form.

The official information on the ISTA Seed Lot Certificate (O.I.C. and G.I.C.) must be provided and signed by the official sampler, otherwise no ISTA Seed Lot Certificate

species				method used			M.C.	analysis number 3
1ST REPLICATE					2ND REPLICATE			
1	2	3	4		5	6	7	8
				number of container				
g	g	g	g	weight of container	g	g	g	g
g	g	g	g	weight of seed	g	g	g	g
g	g	g	g	weight of container plus seed before drying	g	g	g	g
g	g	g	g	weight of container plus seed after drying	g	g	g	g
g	g	g	g	loss of weight	g	g	g	g
%	%	%	%	moisture content	%	%	%	%
				date and initials of analyst				

calculation in case of pre-drying according to the formula:

$$M = S_1 + S_2 - \frac{S_1 \times S_2}{100} = \quad \%$$

S_1 = percentage of moisture lost by pre-drying (stage 1)
S_2 = percentage of moisture lost by the oven method (stage 2)

average moisture content (in case of pre-drying) $\frac{A + B}{2} = \quad \%$

Figure 43. Form 3, Moisture form.
Both replicates and possible retests can be entered on this sheet. For explanation see paragraph 13.4.

may be issued. Where any ISTA Seed Lot Certificate is required, the official sampler not only samples the lot, but also will mark and seal the bags and will forward the sample and the request form to the seed testing station. After receipt the station has to fill in the date of receipt of the sample and the analysis number.

13.3. The cover form (Form 1, figure 42)
As soon as the request form has been received at the station it is put in a cover form together with the other forms. The same analysis number is also put on the sample label.

The cover forms holding the request forms are filed in the administration centre. The other forms are distributed to the sections where the analyses are to be carried out.

The cover form is used to indicate the analyses that must be executed according to the opinion of the person in charge of the seed testing station, based on the national and international regulations and seed laws (sometimes this is not identical with the request of the applicant).

A special place is reserved for the calculated P.L.S. (Pure Live Seed Content): percentage pure seed x percentage normal seedlings/100.

On the right side of the front flap of the cover form the specification of the invoice (specification of charges) is recorded. On the left side the dates of issuing the certificate(s) are noted, and the spare space on the left side can be used for notes concerning the analyses (e.g. retesting necessary).

The invoice is prepared and signed by one person (at the bottom, right side).

Summarising, the cover form is used exclusively by the person(s) responsible for the registration of the sample on receipt, and for the final check of the analysis results just before the preparation of the certificates, as well as for the preparation of the invoice.

13.4. The moisture form (Form 3, figure 43).
The analyses are to be executed in duplicate. From the moisture form the vertical columns 4 and 5 on both sides of the central column are used. The percentage moisture content is calculated as:

$$\frac{\text{loss of weight}}{\text{initial weight of seed}} \times 100$$

When pre-drying, because columns 4 and 5 of the form have already been used for the initial analyses, column 3 has to be used for the analysis of the 50.0 g sample which has to be pre-dried. Thereafter columns 2 and 6 are used for the final moisture content determination. For the second replicate of a pre-drying determination, a second moisture form is needed. Column 4 of this form can be used for the 50.0 g which is to pre-dried, whereafter columns 3 and 5 are used for the final determinations. The two moisture forms should be marked A and B. The moisture content of the pre-dried sample is the average of the results of determinations A and B.

The average moisture content should be written on the upper side of the moisture form, left of the analysis number, and be expressed as a percentage calculated to one decimal place. The method used should be reported on the upper part of the form.

13.5. Germination form (Form 4, figure 44)
The space above the double line is intended for initial information and final results; the space below this line is for the analysis. Before the analysis is started the leading analyst of the germination section indicates on the form the method to be used for the seed that is going to be tested.

After planting, the date of planting and the counting dates are noted. Normally, the seeds are sown in four replicates of 100 seeds each (8 × 50 or 16 × 25). The planted seeds are put in or on a germination apparatus. It is advisable to give each germinator a number and also each tray in each cabinet or each slot of a germination table. The sample can easily be found then. Tray 5 in cabinet 2 will then be indicated by: 2.5, and will be recorded on the germination form, in this way locating the test.

method										
	days	%	number of days	normal seedlings	hard seeds	fresh ungerm. seeds	abnormal seedlings	dead seeds		
counting days	first cnt.		space for additional information						apparatus number	*analysis number* 4
species										

date	number of days	I						II						III						IV						initials of analyst	calculation
		N	H	F	A	D	R	N	H	F	A	D	R	N	H	F	A	D	R	N	H	F	A	D	R		
total																											

space for observations			
	normal	N	%
	hard	H	%
	fresh	F	%
	abn.	A	%
	dead	D	%
	total		%

Figure 44. Form 4. Germination form.
All four replicates including the calculation of the final results can be recorded. For explanation see paragraph 13.5.

In the vertical 'date' column the date of the evaluation is written, e.g.: May 23rd, as 23/5.

In the vertical column 'number of days', the number of days since the date of planting is written. When the seed has been sown on Monday, the next Monday is the seventh full day (not the eighth). Hence, the date of planting is not included. On the horizontal lines the results of the evaluations are written. For each evaluation day one horizontal line can be used.

The categories of evaluated seeds and seedlings are:
N – normal seedlings;
H – hard seeds;
F – fresh ungerminated seeds;
A – abnormal seedlings;
D – dead seeds (e.g. rotten seeds);
R – rest; indicating the number of seeds or seedlings that cannot yet definitely be classified.

An example (see table 2): On the first counting day perhaps in one replicate 40 normal seedlings can be removed (and recorded), but the rest (60) consists of insufficiently developed seedlings for evaluation and (or) ungerminated but not rotten seeds. These 60 are written under R. Rotten seeds and also rotten seedlings are removed and recorded (under D and A respectively) at any time during the test to prevent contamination of the healthy seeds or seedlings. For that reason, frequent inspection of the tests is advised, because removal may be necessary between counting days.

On the next counting day these 60 seeds or seedlings have to be evaluated anew. Now perhaps 30 seedlings are normal, three seedlings decayed and five seeds clearly rotten. The rest to be written under R is now 22. Apart from rotten seedlings no abnormal seedlings are removed from the test until the final count, because this facilitates checking of the type of abnormalities by the leading analyst.

On the last counting day the evaluation has to be completed. Now the 22 seeds and seedlings of column R must be subdivided as far as possible. Let us suppose that there are now:

2 normal seedlings	(N)
2 hard seeds	(H)
3 fresh ungerminated seeds	(F)
12 abnormal seedlings	(A)
3 rotten seeds	(D)

For each replicate the results must be stated in the appropriate vertical column I, II, III or IV.

The results of the replicate of the sample are scored as follows:

Table 2.

Date	Number of days	I					
		N	H	F	A	D	R
23/5	4	40	–	–	–	–	60
27/5	8	30	–	–	3	5	22
31/5	12	2	2	3	12	3	–
		72	2	3	15	8	–
				TOTAL: 100			

The same recording is done for the other replicates. For the final calculation of the average values the right bottom side is used. As the germination test is performed in four replicates, the averages are often not whole figures; 0.25 is rounded downwards and 0.50 and 0.75 are rounded upwards. Because of the rounding off, the total percentages of all categories together will sometimes deviate from 100%, in which case a correction is necessary. This correction is never executed in the category N.

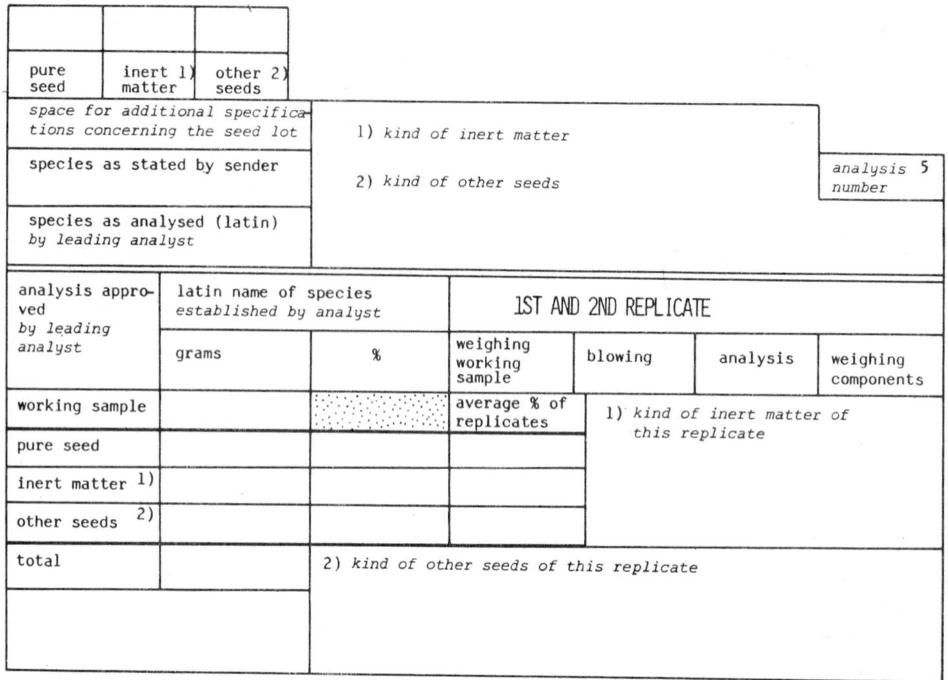
pure seed
inert matter 1)
other seeds 2)
space for additional specifications concerning the seed lot
1) kind of inert matter
species as stated by sender
2) kind of other seeds
analysis number
5
species as analysed (latin) by leading analyst
analysis approved by leading analyst
latin name of species established by analyst
1ST AND 2ND REPLICATE
grams
%
weighing working sample
blowing
analysis
weighing components
working sample
average % of replicates
1) kind of inert matter of this replicate
pure seed
inert matter 1)
other seeds 2)
total
2) kind of other seeds of this replicate

Figure 45. Form 5, Purity form.
One replicate is recorded on one sheet; if tests are carried out in duplicate two forms are needed. For explanation see paragraph 13.6.

As an example, when te average of four replicates is:

	after rounding off:	the correction then will be:
N = 70.50%	N = 71%	N = 71%
H = 12.50%	H = 13%	H = 12%
F = 2.50%	F = 3%	F = 3%
A = 6.25%	A = 6%	A = 6%
D = 8.25%	D = 8%	D = 8%
Total = 100.00%	Total = 101%	Total = 100%

The form can also, with some improvising, be used for the recording of tetrazolium results.

13.6. Purity form (Form 5, figure 45)

Like the other forms the purity form has two main sections separated by a double line. The lower section is for the analysis, the upper section is for the final conclusions and additional information.

A purity test should be executed in duplicate, two subsamples (half working samples) being independently drawn and tested. The two tests are referred to as the '1st replicate'

and the '2nd replicate'. Two purity forms should therefore be prepared, one to be filled with the results of the '2nd replicate', the other to be filled with the results of the '1st replicate' together with the average of the two. In that way the results obtained by one analyst cannot bias the other, the two sets of results being compared after the percentages have been calculated. The leading analyst of the purity section, when receiving a submitted sample with its duplicate purity analysis forms, designates two analysts to carry out the analyses and to draw the duplicate half working samples.
Each analyst:
- checks whether the analysis number on the label of the submitted sample agrees with the number on the form;
- writes the analysis number, species name, '1st' or '2nd' on the working sample container, and signs it;
- takes a working sample (using the procedure as outlined in paragraph 8.1), determines its weight, writes the outcome at 'working sample' at the top of the 'grams' column of the form, and writes the date and signs at 'weighing working sample';
- (if the sample requires fractionation by blowing, either as an aid or in pursuance of a prescription) subjects the working sample to blowing and encloses the fractions in separate containers;
- writes the blowing speed (= manometer reading, or valve opening reading) on each container;
- writes the blowing speed, the date and signs at 'blowing' on the form;
- takes the fractions and the form to a working table.

At the working table, the same or another analyst:
- subjects the subsample (whether or not fractionated) to a purity test;
- writes the kind of inert matter behind 1) and the scientific (Latin) names of the other seeds behind 2);
- writes the scientific name of the pure seed at 'Latin name of species (established by analyst)';
- notes the date and signs in the 'analysis' space;
- has data checked by the leading analyst;
- weighs the components in the weighing unit.

The leading analyst is responsible for any results that are submitted for reporting and signs and dates the form checking the name of the pure seed species by writing it under 'species as analysed' before transferring the form to the office if the calculation is performed by the administration centre.
In the office the clerk:
- calculates the component percentages (to two decimal places);
- compares the two series of percentages that result from the two replicates, and compares the difference with the prescribed table of tolerances;
- (if the difference between comparable percentages exceeds tolerance) takes two blank purity analysis forms, writes the analysis number in its upper right corner and orders an additional purity test;

analysis results							
							analysis number 6
species as stated by sender	*other determinations*			Cuscuta	Orobanche	other species by number	1000 seed weight
species as analysed (latin) *by leading analyst*	1ST/2ND REPLICATE			grams	analysis	weighing	analysis approved
space for results of one replicate							

Figure 46, Extra form.
For recording special tests. For explanation see paragraph 13.7.

- (if the difference between comparable percentages is within tolerance) writes the average percentage of the test in column 'average of replicates' of one of the two forms;
- rounds off these percentages to one decimal place;
- corrects the percentages such that their total is 100.0%, by correcting the inert matter fraction only (impurities that occur to an extent of less than 0.05% are reported as 'trace', and are not included in this calculation);
- copies the corrected percentages at the head of the first form;
- copies (in the middle part of the head of the form), at '1' and '2' the data from the '1st replicate' and the '2nd replicate' of 'kind of inert matter' and 'kind of other seeds', by adding them up. Then, the data on the form are ready to be reported on a report form (certificate).

13.7. Extra form (Form 6, figure 46)

This form serves all kinds of special determinations that normally will be reported under the heading 'Other Determinations' of an ISTA certificate. Such tests include: other seeds by number (number count test), 1000 seeds weight, hectolitre or bushel weight, verification of species or cultivar, seed health tests, percentage of sprouted grain tests, percentage of cracked seeds tests, etc., etc.

14. Equipment

*14.1. Recommended furnishing and equipment**

The equipment is arranged according to the chapters and paragraphs used in this paper. For total number per item needed (both for the 2000 and the 5000 laboratory), refer to paragraph 14.2.

5. Sample reception

- Rack for blank forms: 4 shelves, 30 × 220 cm, fixed to the wall at height of 60 cm, 90 cm, 120 cm and 150 cm.
- Table: top 75 × 400 cm, at 95 cm above floor level; two additional shelves underneath, 65 × 400 cm, at 20 cm and 50 cm above floor level; heavy duty.
- Numbering machine, dating stamp, stock of forms.
- Chairs in the hall to accommodate visitors.

6. Administration centre (office)

- Filing cabinet; a standard type, 65 cm deep, 54 cm wide, 138 cm high, with six drawers, each with a partition to hold two rows of forms produced in a single season.
- Desk (for the clerk/typist) with one set of drawers.
- Desk (for the superintendent) with two sets of drawers.
- Open tray, for temporarily filing analysis forms: inside dimension 22 × 45 × depth 10 cm, with adjustable inclined partitions that prevent the forms from slipping.
- Simple electronic calculating machine, for percentages and other calculations.
- Typewriter, telephone, stationery, stock of report forms (or certificates).
- Chairs.

7. Moisture unit

7.2. Oven method (figure 5)

- Aluminium containers, 6 cm diameter, 3 cm high with loose fitting lids. Supplier 1: Centraal Magazijn, Industrieweg 5, 3433 NL Nieuwegein, Netherlands; Supplier 2: H. B. Selby & Co., 48 North Terrace, Adelaide, South Australia 5000, Australia.
- Drying oven. Recommended oven T 5042 E + additional thermometer insertible in top. Supplier: Heraeus GmbH, P.O. Box 169, 645 Hanau, Germany F.R.
- Balance. Recommended type of balance is: 'precision balance 2' (see below under 8.2.).
- Grinding mill. Recommended Cemotec. Supplier: Tecator AB, P.O. Box 70, S-263 01 Höganäs, Sweden.
- Sieves. 0.5, 1.0 and 2.0 mm mesh + receptacle. Supplier: Duintjer & Zn., P.O. Box 1, 9640 AA Veendam, Netherlands.
- Desiccator ± 20 cm diameter, with thick metal or porcelain plate and blue silicagel. Readily available through any dealer of laboratory supplies.
- Pair of tongs. Model 1309 B3. Supplier: Tamson, P.O. Box 208, 2700 AE Zoetermeer, Netherlands.

7.3. Quick methods

7.3.1. Recommended: Ultra X 70 (figure 6). Supplier: A. Gronert, P.O. Box 167, D 4937 Lage (Kachtenhausen), Germany F.R. Alternative 1: Cenco moisture balance 26680–817. Supplier: Breda Scientific, P.O. Box 3336, 4800 DH Breda, Netherlands. Alternative 2: Mettler LP 15 combined with balance PC 440 and application input unit GC 301. Supplier: Mettler Instruments, CH-8606 Greifensee, Switzerland.

* The recommendation of particular models and listing of their suppliers by the authors does not necessarily indicate approval by the International Seed Testing Association.

7.3.2. Recommended: Dickey-John DjGMT Grain Moisture Tester (Figure 7). Portable, battery operated, digital read-out. Supplier: Dickey-John, P.O. Box 10, Auburn, IL 62615, USA. Working on the same principle, portable, battery operated, direct read-out: Cera-Tester. Supplier: A/S N. Foss Electric, 69 Slangerupgade, DK- 3400 Hillerød, Denmark.

8. Purity section

8.1. Subsampling unit

- Divider (figure 8). Recommended: riffle type divider (soil divider), Model INRA. Supplier: Tripette & Renaud, 39 Rue J.-J. Rousseau, 75038 Paris Cedex 01, France.
- Aluminium containers (figure 17). Supplier: Centraal Magazijn, Industrieweg 5, 3433 NL Nieuwegein, Netherlands. Alternative: paper bags, various sizes.
- Shallow tray, straight-edged spoons and spatula (figure 9), various sizes. These items are to be made locally.

8.2. Weighing unit

- Balances. Although mechanical balances have our preference, mainly because they are cheap, we advise electronic balances. Good mechanical balances are no longer available, or will not be available in the near future. Because the after-sales service is very important, in many countries the choice of make will be limited to two main producers of laboratory balances: Mettler and Sartorius. Every year new improved models are developed and therefore advice from our side must be limited to the type of balance (see table 1 of paragraph 8.2 in main text). Both laboratories need an analytical balance (figure 10) and a precision balance 1, but if possible also a precision balance 2 ('intermediate' type) could be purchased and will prove to be very useful. For addresses of local agents write to: Supplier 1: Mettler Instrumente AG, CH-8606 Greifensee, Switzerland; Supplier 2: Sartorius GmbH, P.O. Box 19, D-3400 Göttingen, Germany F.R.
- Weighing table (figure 11). Ready-made tables can be purchased but very often one will have to make one oneself, which is not difficult. The table consists of a heavy (metal or brick) base and a thick stone slab (± 8 cm thick). The size of the table depends on the number of balances available and the place where they are needed. Per balance one will need a table surface of ± 75 × 75 cm. The analytical balance should have its own table. The slab should not be laid directly onto the base but on rubber blocks. In this way the heavy table top is more or less independent from the vibrations that still might occur in the base. To limit these vibrations in the base it is essential that it stands on a solid concrete floor (if available) or that it is mounted to a solid stone or concrete wall (if available). If neither of these are available accurate weighing will be impossible.

8.3. Blowing unit

- Seed blower (figure 12). Recommended: Micro Blower type 35. Supplier: Contab Instrumentation, Klisätravägen 25 A, S-130 12 Älta (Stockholm), Sweden. Alternative 1 (for fine grasses): Ottawa blower. Supplier: Astell Hearson, 172 Brown Hill Road, Catford, London SE6 2DL, United Kingdom. Alternative 2 (for coarse separations of the larger grasses): South Dakota Blower. Supplier: Seedburo, 1022 West Jackson Blvd., Chicago IL 60607, USA.

8.4. Purity unit

- Working tables (figure 13). Although any desk that is standing firmly on the ground is suitable for a purity test, it is advantageous to have a special table constructed. This table is specifically designed for the purpose. A major advantage is that the elbows have a good support; this is very important because the analyst is usually sitting at the table for several hours every day. A good support of the elbow prevents tiredness and in the long term serious physical complaints. A diaphanoscope can be made in the table by cutting out a 4.7 × 4.7 cm rectangle, covering it by a glass plate that fits flush into the table, and mounting a small 20 W (28 V) microscope lamp underneath. In that case also an additional funnel fitted into the table is recommended. Freestanding diaphanoscopes are available, that can be used on any table: Illuminant Magnifying Glass with polarizing filters. Supplier: Tripette & Renaud, 39 Rue J.-J. Rousseau, 75038 Paris Cedex 01, France.

- Incident light. Recommended: Luxo FL-101A. Supplier: write for local agent to: Jacobson, P.O. Box 60, Oslo-6, Norway.
- Working board (figure 16). The board (30 × 50 cm) may consist of hardboard, hardboard sandwiched between dull hard plastic of a dark colour, or of sanded (dull) black glass. In any case it should be of a hard, smooth, non-static material.
- Binocular microscope (figure 16). Most binocular microscopes have a range of magnifications. If variable ('zoom') then a combination of lenses should be chosen to get a range from about 6–25 ×. If two single magnifications are possible, then 6 × and 16 × are the best combination. If only one magnification is available then 16 × is the best choice. In that case an additional (screw-on) lens of 0.5 × is often available, which will give the possibility of 8 × magnification too. Attention should be given to the stand (see main text).
- Magnifier. 3, 4 or 5 × magnification on stand. Recommended (4 × magnification): type 15900. Supplier: Breukhoven, P.O. Box 6044, 3002 AA Rotterdam, Netherlands.
- Metal spatula, tweezers and scalpel needle (figure 14). Can easily be made locally. Tweezers of the recommended type are also: 1068 B 10. Supplier: Tamson, P.O. Box 208, 2700 AE Zoetermeer, Netherlands.
- Seed collection (figures 15 and 16). A small quick-reference seed collection (figure 16) can be made by using small glass test tubes inserted in a wooden block that is provided with holes. All analysts should prepare such a collection for their own use. For the main seed collection more provisions are needed: recommended (figure 15) are metal units with 48 plastic drawers. In each drawer up to eight glass test tubes (maximum length 12 cm) can be accommodated per unit resulting in a collection of up to 384 samples. Recommended type: 48A. Supplier: write for local agent to: Raaco A.S.E., 4800 Nykøbing/Falster, Denmark.
- Half funnel. Will have to be home-made.
- Watch glasses. Available through any dealer in laboratory supplies.

9. Germination section

9.2. Germination rooms

- Germination rooms. Prefabricated with well insulated panels with varying thickness of the insulation material and air-conditioning. Supplier: Zephyr, P.O. Box 507, 2700 AM Zoetermeer, Netherlands.

9.3. Germination cabinets

Depending on the circumstances, the kinds of seed to be tested and the financial possibilities, four types can be recommended:

- Inventum 'wet' cabinet – with light (DK-10, figure 21). The cabinet serves all purposes but is especially suitable for carrying out tests with the various paper substrates. No covering of the filter paper is needed, since there is no danger of drying out. Although not strictly necessary, the tests are usually placed on a moist filter paper underlay. Water flowing along the inner sidewalls provides a humidity of 100% and also regulates the temperature: constant and alternating temperatures. Supplier: Inventum, P.O. Box 4, 3720 AA Bilthoven, Netherlands.
- Zephyr 'wet' cabinet – without light. With constant and alternating temperature conditioning. The humidity and temperature are regulated by water delivered through a spray from a distributor pipe in the space between the stainless steel linings which wets both inner and outer lining. Tests do not dry out. The cabinet can be fully compared with the Inventum cabinet but is without the light facility, which limits its possibilities. Supplier: Zephyr, P.O. Box 507, 2700 AM Zoetermeer, Netherlands.
- Zephyr 'dry' cabinet – with and without light (ZGM-E, figure 22). With constant and alternating temperature conditioning. Tests have to be carried out in containers with covering. Supplier: Zephyr, for address see above.
- Navep 'dry' cabinet. Several types: with and without light, with only constant temperature conditioning or also with alternating temperature conditioning. Light is supplied by four loose cold fluorescent tubes of 8 W which can be placed at any position in the cabinet. As with the Zephyr cabinet, homogeneous temperature distribution is achieved by a ventilator for forced air circulation. Supplier: Navep, P.O. Box 256, 9700 AG Groningen, Netherlands.

9.4. Germination tables

Two alternatives are recommended:

- Inventum table (figure 23). The germination plate of stainless steel on which the germination beds are placed consists of eight parallel, rectangular tubes of stainless steel through which temperature controlled water is pumped to achieve the required temperature. The tubes are separated by a 7 mm gap. The level of the water in the tank underneath the germination plate is automatically controlled and can be changed (cleaned) by pressing a button. The whole plate can be jacked up to clean the tank thoroughly. The plate can hold 144 or 207 round beds of 10 cm and 8 cm diameter respectively. Supplier: Inventum, P.O. Box 4, 3720 AA Bilthoven, Netherlands.
- Zephyr table. The table has a removable plate of stainless steel. Underneath a copper pipe system is fitted through which the temperature regulated water is pumped to achieve the required germination temperature. The plate has as standard 176 slots cut into it on which beds of 8 cm diameter are placed and through which the paper wicks hang down into the stainless steel water bath underneath. Supplier: Zephyr, P.O. Box 507, 2700 AM Zoetermeer, Netherlands.

Note: All germination equipment is normally delivered for 220 V–50 Hz power supply. However, the power supply of all apparatus can be adapted upon request.
Light (if supplied) and temperature are automatically operated with electronic temperature regulators and timers.

9.5. Counting devices

9.5.1. Vacuum counting heads (figure 25)

For requirements see main text. Supplier: Ames Powercount Co. & El Drechson Co., Brookings, SD 57006, USA. The counting heads should be used in combination with a central vacuum circuit or a powerfull vacuum cleaner that is not susceptible to over-heating. Recommended: Nilfisk GS 80. Supplier: write for local agent to Fisker & Nielsen A/S, Peter Bangsvej 30, DK-2000 Copenhagen F, Denmark.

9.5.2. Counting boards (figures 26 and 30)

The board that is depicted in figure 26 is especially designed for the Bema germination container for sand tests (see below). It may be possible to have it locally made. Technical plans can be provided by the authors.

9.8. Further needs

- Rakes and scrapers. Can be made using figures 27 and 14.c respectively. The scraper fits into the box shown in figure 39.
- Containers with transparent cover for sand tests (figure 39). Recommended: Bema kiembakje. Supplier: Bema, Deltastraat 14, 4301 RC Zierikzee, Netherlands.
- Containers for paper tests (figure 40). Recommended: box 21 × 15 × 3 cm. Supplier: D.B.P. Plastics, Terbekehof 25–29, Wilrijk, Belgium.
- Bell jars (figure 24). 8 cm diameter. Supplier: Tamson, P.O. Box 208, 2700 AE Zoetermeer, Netherlands. 10 cm diameter. Supplier: Leithen Valley Plastics, Leithen Road, Innerleithen, Peebleshire, Scotland.
- Filter paper. Circles of 8 cm or 10 cm diameter and 0.6 mm thickness (code T 300), and also the rectangular sheets (code ZH 1220, ZH 1224) are supplied by: Schut & Zn., P.O. Box 1, 6866 ZG Heelsum, Netherlands. Pleated filter paper (code 3014) and the coverslip that is also needed (code 0858) are supplied by: Schleicher & Schüll, P.O. Box 4, D-3354 Dassel, Germany F.R.
- Pre-drying oven. The oven of the moisture department may be used. If an additional oven is needed, refer to 7.2 for supplier.
- Mixer. e.g. Lescha VM 130 N. Supplier: Gehabouw, P.O. Box 246, 3900 AE Veenendaal, Netherlands.
- Washing apparatus for beet seed. Supplier: J. Volkers Jr., ARVO, Bovenweg 41, 1834 CB Sint Pancras, Netherlands.

14.2. Survey of total equipment needed

	Recommended quantity	
	2000 laboratory	5000 laboratory
Moisture		
aluminium container	30	50
oven	1	1
grinding mill	1	1
set of sieves	1	1
desiccator	2	2
pair of tongs	1	1
laboratory chair	1	1
quick moisture tester	optional	optional
Purity		
divider	1	1
analytical balance	1	1
precision balance (capacity 1000 g)	1	1
weighing table	2	2
seed blower[1]	1	1
working table	2	4
built-in diaphanoscope[1]	2	4
or: freestanding table model	1	2
incident light source	3	5
binocular microscope	1	2
magnifier	3	5
cupboard for seed collection	1	1
small aluminium container[2]	1000	2000
large aluminium container[2]	500	1000
large paper bag (12 × 24 cm)[3]	2000	5000
laboratory chairs	2	4

In addition small tools are needed: shallow trays, straight-edged spoons and spatulas, tweezers, scalpel needles, working boards, half funnels, watch glasses, test tubes for seed collection (see main text).

[1] Only where grasses have to be tested.
[2] Paper bags can be used instead (suggested size: 6 × 9 cm), but have the disadvantage that they can not be re-used because of danger of contamination of one sample with the other.
[3] To accommodate the working sample and the pure seed in cases where the amount of seed to be examined does not fit into the aluminium containers (e.g. with pulses, cereals or in the case of number count tests).

	Recommended quantity	
	2000 laboratory	5000 laboratory
Germination		
germination room	1	2
germination cabinet with light	1	1
germination table	1	1
vacuum counting device with two heads (for cereals and brassica – sized seeds)	optional	1
counting board	4	4
containers with transparent cover for sand tests[1]	600	1500
containers for paper tests[1]	50	250
bell jars[2]	200	200
pre-drying oven	optional	optional
mixer	optional	optional
washing apparatus[3]	optional	optional
domestic refrigerator	1	1
hot water boiler	1	1
Viability		
binocular microscope	–	1
small utensils: magnifier, small glass containers, tweezers, lancet knives, needles.		
Administration + sample reception		
rack for blank forms	1	1
table + shelves	1	1
numbering machine	–	1
dating stamp	1	1
chairs (visitors)	4	2
filing cabinet	1	1
desk	2	2
open tray for forms	1	2
calculator	1	1
typewriter	1	2
chairs	3	4

In addition small tools are needed: rakes, scrapers, tweezers, spatulas, magnifiers, watering bottles, sieves (to separate seeds from sand). Refer to main text.

[1] Numbers are approximate and will depend on the species to be tested.
[2] Size depends on the size of the filter paper circles used (8 cm or 10 cm diameter).
[3] Only when many beet seeds have to be tested.

15. List of books and journals

A basis for a good seed testing library.

General

International rules for seed testing (Rules + Annexes)
Seed Science and Technology 4 (1), 1976.
German edition: Seed Science and Technology 4 (3), 1976.
French edition: Seed Science and Technology 4 (4), 1976.
ISTA Secretariat*.

Amendments to international rules for seed testing 1976 – composite version of 1977 and 1980 admendments
S. R. Cooper, ISTA Secretariat*, 1981.

Manual for testing agricultural and horticultural seeds
Agriculture handbook No. 30. USDA, Washington DC, USA, 1952.

Survey of equipment and supplies for seed testing
ISTA Secretariat, 1982.

Seed technology
R.L. Agrawal. Oxford & IBH Publishing Co., 66 Janpath, New Delhi, India, 1980.

An introduction to seed technology
J. R. Thomson. Leonard Hill, Glasgow, UK, 1979.

Principles of seed science and technology
L. O. Copeland. Burgess, Minneapolis, USA, 1976.

Seeds, the yearbook of agriculture
USDA, Washington DC, USA, 1961.

Diseases, pests and weeds in tropical crops
J. Kranz, H. Schmutterer and W. Koch. Parey, Berlin, German Federal Republic, 1977.
(Also in German).

Handbook on seed health testing
ISTA Secretariat*, 1979.

An introduction to the botany of tropical crops
2nd. edition. L. S. Cobley and W. M. Steele. Longman, London, UK, 1976.

Agricultural and horticultural seeds – their production, control and distribution
Plant Production and Protection series No. 12.
FAO, Rome, Italy, 1961.
(also in French: Semences agricoles et horticoles; and in Spanish: Las semillas agricolas y horticolas)

Improved seed production – a manual on the formulation, implementation and evaluation of seed programmes and projects
Plant Production and Protection series No. 15.
FAO, Rome, Italy, 1978.
(also in French: Production de semences améliorées; in Spanish: Mejoramento de la produccion de semillas; and in Chinese)

Seed Production
P. D. Hebblethwaite. Butterworths, London, UK, 1980.

Seed Science and Technology (journal)
ISTA Secretariat*

Journal of Seed Technology (journal)
AOSA*

Seed abstracts (journal)
CAB, Farnham Royal, Slough, UK

Advances in Research and Technology of Seeds (journal)
Pudoc, Wageningen, Netherlands

Purity

ISTA List of stabilised plant names
ISTA Secretariat*, 1966.

Supplementary list of the 'ISTA List of stabilised plant names 1966' of plant names stabilised by ISTA since 1966 until the end of 1980
ISTA Secretariat*, 1981.

A multilingual glossary of common plant names
1. *Field crops, grasses and vegetables,* 2nd edition. ISTA Secretariat*, 1982.
2. *Trees,* 1st edition. ISTA Secretariat*, 1971.
(also: Proceedings of the International Seed Testing Association, 37 (5))

Identification of crop and weed seeds
A. F. Musil. Handbook No. 219. USDA, Washington DC, USA, 1963. (reprinted by: Castle House, Turnbridge Wells, UK, 1980)

Common weeds from Iran, Turkey, the Near-East and North Africa
F. Bischof. Schriftenreihe Nr. 49. GTZ, Eschborn, German Federal Republic, 1978.

Handbook on Pure Seed Definitions
E. M. Felfoldi. ISTA Secretariat*, in preparation.

Local floras

Germination

Handbook for seedling evaluation
ISTA Secretariat*, 1979.

Manuel para evaluacion de plantulas en analisis de germinacion
(translation of previous handbook).
Inst. Nac. de Semillas y Plantas de Vivero, Carretera de la Coruña, KM. 7.500, Madrid – 35, Spain, 1980.

SEEDLAB 2000–5000

Seed biology
Vols. 1, 2, 3. T. T. Kozlowski, Academic Press, New York, USA, 1972.

Seed ecology
W. Heydecker. Butterworths, London, UK, 1973.

The germination of seeds
3rd. edition. A. M. Mayer and A. Poljakoff-Mayber. Pergamon, Oxford, UK, 1982.

Physiology and biochemistry of seeds in relation to germination
Vols. 1 & 2. J. D. Bewley and M. Black. Springer, Berlin, German Federal Republic, 1978.

Physiology of deep dormancy in seeds
M. G. Nikolaeva. Israel Program for Scientific Translations, Jerusalem, Israel, 1969.

Seeds of woody plants in the United States
Handbook no. 450. USDA, Washington DC, USA, 1974.

Viability

Tetrazolium testing handbook for agricultural seeds
D. F. Grabe. Handbook on seed testing – Contribution No. 29. AOSA*, 1970.

Tetrazolium Testing Handbook
ISTA Secretariat*, in preparation.

Storage

Viability of seeds
E. H. Roberts. Chapman and Hall, London, UK, 1972.

Recalcitrant crop seeds
H. F. Chin and E. H. Roberts. Tropical Press, Kuala Lumpur, Malaysia, 1980.

Principles and practices of seed storage
O. L. Justice and L. H. Bass. Handbook no. 506. USDA, Washington DC, USA, 1978. (also: Castle House, Turnbridge Wells, UK, 1979)

Seed preservation and longevity
L. V. Barton. Leonard Hill, London, UK, 1961.

* Present address of:

ISTA Secretariat
Reckenholz, P.O. Box 412
CH–8046 Zürich, Switzerland

AOSA
Charles C. Baskin
Mississippi Extension Service
P.O. Box 5267
Mississippi State University
Mississippi State, MS 39762
USA